Quantum Computing Algorithms

Discover how a little math goes a long way

Barry Burd

BIRMINGHAM—MUMBAI

Quantum Computing Algorithms

Associate Group Product Manager: Gebin George
Publishing Product Manager: Kunal Sawant
Senior Content Development Editor: Rosal Colaco
Technical Editor: Jubit Pincy
Copy Editor: Safis Editing
Associate Project Manager: Manisha Singh
Proofreader: Safis Editing
Indexer: Pratik Shirodkar
Production Designer: Shyam Sundar Korumilli
Business Development Executive: Kriti Sharma
DevRel Marketing Coordinator: Sonia Chauhan

First published: August 2023

Production reference: 1310823

Published by Packt Publishing Ltd.
Grosvenor House
11 St Paul's Square
Birmingham
B3 1RB, UK

ISBN 978-1-80461-737-3

www.packtpub.com

For

Abram and Katie, Benjamin and Jennie, Sam and Ruth, Harriet, Sam, and Jennie, Abram and Jennie,

- *Barry Burd*

Contributors

About the author

Barry Burd received a master's degree in computer science at Rutgers University and a Ph.D. in mathematics at the University of Illinois. As a teaching assistant in Champaign–Urbana, Illinois, he was elected five times to the university-wide *List of Teachers Ranked as Excellent by Their Students*.

Since 1980, Dr. Burd has been a professor in the department of mathematics and computer science at Drew University in Madison, New Jersey. He has spoken at conferences in the United States, Europe, Australia, and Asia. In 2020, he was honored to be named a Java Champion.

Dr. Burd lives in Madison, New Jersey, USA, where he spends most of his waking hours in front of a computer screen.

I extend thanks (in reverse alphabetical order) to the following people for their help and support: George Zipperlen, James Weaver, Johan Vos, Ron Schreiner, Naria Rush, Anna Naden, Bob Miller, Terrill Frantz, Dhruv Erry, and Bambordé Baldé.

About the reviewers

Rakshit Jain is an IBM Certified Associate Quantum Developer who is very passionate about operating systems, quantum computing, and cyber security (white hat, obviously). He's currently in the penultimate year of his studies at the Hong Kong Polytechnic University, where he leads the Google Developer Student Club as the vice president. Driven by curiosity, he experiments with (and breaks) any bleeding-edge technology he can get his hands on. When he's not writing scripts, he is DJing at events or climbing mountains.

Ron Schreiner has been fortunate to have worked in a broad range of technical disciplines during his career. Some of these areas include digital hardware design, embedded CPUs, silicon functional architecture (VLSI), computer compiler development, image/live video processing, and others. As a provider of medical software, he has recently delved into machine learning and most recently, quantum computing with an emphasis on quantum machine learning. Outside of the technical arena, Ron is a licensed pilot, certified open-water SCUBA diver, bicyclist, woodworker, and a once-in-a-while artist.

Table of Contents

3

Math for Qubits and Quantum Gates   61

4

Qubit Conspiracy Theories   89

Part 2: Making Qubits Work for You

5

6

Part 3: Quantum Computing Algorithms

7

8

Grover's Algorithm   197

9

Shor's Algorithm   235

Part 4: Beyond Gate-Based Quantum Computing

10

Some Other Directions for Quantum Computing   279

Assessments   289

Index   311

Other Books You May Enjoy   320

Preface

Several years ago, in the waiting room of a physician's office, I picked up a copy of the magazine *Nature*. One of the issue's articles lit a fire under me and led me to the study of quantum computing. At first, I was perplexed. How can this model of computing be so different from the kind of computing we know and love? Why do these strange circuits yield correct answers to mathematical problems?

Being a teacher, I knew that the best way to learn something was to try explaining it to someone else. So I decided to offer a one-semester course for university students. A typical quantum computing introduction involves programming, physics, and abstract mathematics. I wanted my course to attract students from many disciplines, and frankly, I didn't want the course to be canceled for lack of enrollment. To make this work, I had to carve out a section of the subject that required a carefully measured amount of background knowledge. I wanted students to perform some calculations. But first and foremost, I wanted them to understand *why* certain quantum computing algorithms work. What does a particular quantum circuit have to do with the problem that they're trying to solve?

After several false starts, I landed on a plan to focus on matrix operations. Students learned about matrix multiplication and the tensor product. They programmed quantum computers on IBM's cloud and performed routine calculations to verify the results. Through in-class explorations, they developed insights into the way quantum computing algorithms work. Best of all, they enjoyed the course.

In 2022, the people at Packt asked me to write a book on a particular topic that didn't interest me. So, I turned the tables on them and proposed a quantum computing book. They took the bait, and the result is the book that you're reading right now. To be clear, this isn't a textbook, nor is it a scholarly volume. It's not a comprehensive guide of any kind. It's my humble attempt to describe quantum computing algorithms in some detail using only a modest dose of mathematics.

I summarize my approach to the subject in the following bullets:

- I introduce matrix operations in *Chapter 1*. Throughout the book, I use matrices to describe basic ideas about quantum computing. I avoid the abstractions that linear algebra requires.

- I favor examples with actual numbers over formulas and equations.

- When in doubt, I show all the steps in each of the calculations. If there's too much detail for you, just skip over some parts.

- My most important goal is to help you form intuitions about the way quantum computing algorithms work. I use pictures, analogies, and some silly stories to help you understand *why* the algorithms do what they do. (Some quantum computing algorithms are quite challenging in this respect, but I try as best I can.)

This approach has some built-in limitations. If you hated math when you went to school and you never want to see a math problem ever again, this book might not be for you. On the flip side, you may have already studied linear algebra with its vector spaces and eigenvalues. If so, you'll learn less from this book than you would from a more advanced introduction. (Without abstract mathematics, we're limited in what we can cover.)

But I've developed this book's material during three offerings of an introductory college course. And, with each offering, the results were outstanding. Students learned a lot about quantum computing and recommended the course to their friends. I have three semesters of exams and course evaluations to prove all this! So, I hope you keep reading. If all goes as planned, you'll find this material to be interesting, enlightening, and maybe even entertaining!

Who this book is for

This book is for anyone interested in new ways to solve problems. You may have read some news articles about quantum computing and want to dig a bit deeper. Or, you may have explored some technical literature and found it to be too dense. You may be thinking about your career choices and want to investigate quantum computing.

To appreciate this book, all you need is some exposure to secondary school algebra and some exposure to computer programming. You should be comfortable learning about scientific concepts.

What this book covers

Introduction to Quantum Computing, provides an overview of quantum computing and defines some important terms.

Chapter 1, New Ways to Think about Bits, shows how matrix operations can be used to describe logic gates in classical computing. This chapter also introduces Jupyter notebooks as an environment for writing and running computer programs.

Chapter 2, What Is a Qubit?, describes superposition – a strange characteristic of quantum entities that's vastly different from the behavior of classical bits. This chapter also introduces Qiskit – IBM's quantum computing software development kit.

Chapter 3, Math for Qubits and Quantum Gates, provides a way of using matrices to represent qubits' states.

Chapter 4, Qubit Conspiracy Theories, introduces entanglement using multi-qubit gates.

Chapter 5, A Fanciful Tale about Cryptography, shows how quantum superposition can be used to ensure secure communication between two parties.

Chapter 6, Quantum Networking and Teleportation, describes a way in which one qubit's state can be transferred to another qubit. This is an essential element in the transmission of quantum information from one place to another.

Chapter 7, Deutsch's Algorithm, explains how a quantum computer can use superposition and entanglement to solve a simple problem in one step.

Chapter 8, Grover's Algorithm, shows how a quantum computer can search through a list without examining each of the list's elements.

Chapter 9, Shor's Algorithm, describes the way quantum computers will be able to break some of our commonly used cryptographic schemes.

Chapter 10, Some Other Directions for Quantum Computing, describes some alternative approaches to quantum computing.

Assessments contains answers for questions from all the chapters.

To get the most out of this book

You should be comfortable with basic algebra and should have written some computer code. Advanced knowledge in either of these fields is not necessary. No prior understanding of quantum physics is needed. You should be able to think logically about scientific concepts.

This book's code examples use a very small part of the Python programming language. If you have rudimentary skills with Java, C/C++, Visual Basic, or any related language, you can understand the book's examples. If you run into a Python feature you don't already know about, you can learn about it by searching the web.

Software/hardware covered in the book	System requirements
Python	A modern web browser running on any operating system.
IBM Qiskit	
Quantum computers on the IBM cloud	

If you are using the digital version of this book, we advise you to type the code yourself or access the code from the book's GitHub repository (a link is available in the next section). Doing so will help you avoid any potential errors related to the copying and pasting of code.

You can run each of this book's examples on a web page in the **IBM Quantum Lab**. As long as you have a web browser, no other software is required.

But you don't have to rely on IBM's website. If you're willing to follow some setup instructions, you can download Qiskit and run each example on your own computer. Your own laptop can simulate the behavior of a very small quantum computer. In this book, we emphasize the use of IBM's website for running code, but the local execution alternative is always available.

Download the example code files

You can download the example code files for this book from GitHub at https://github.com/PacktPublishing/Quantum-Computing-Algorithms. If there's an update to the code, it will be updated in the GitHub repository.

We also have other code bundles from our rich catalog of books and videos available at https://github.com/PacktPublishing/. Check them out!

Conventions used

There are a number of text conventions used throughout this book.

Code in text: Indicates code words in text, database table names, folder names, filenames, file extensions, pathnames, dummy URLs, user input, and Twitter handles. Here is an example: "In the second cell, type print(x), and then press *Shift+Enter.*"

A block of code is set as follows:

```
circ = QuantumCircuit(2, 1)
circ.h(1)
circ = get_oracle(circ, function)
circ.measure(0, 0)
display(circ.draw('latex'))
```

When we wish to draw your attention to a particular part of a code block, the relevant lines or items are set in bold:

```
device = provider.get_backend('ibmq_qasm_simulator')
job = execute(circuit, backend=device, shots=1000)
print(job.job_id())
result = job.result()
counts = result.get_counts(circuit)$
```

Any command-line input or output is written as follows:

```
$ mkdir quantum
$ cd quantum
```

Bold: Indicates a new term, an important word, or words that you see onscreen. For instance, words in menus or dialog boxes appear in **bold**. Here is an example: "In *Figure 8.30*, the **alice and bob** qubit checks to see whether both Alice and Bob get their wishes."

> **Tips or important notes**
> Appear like this.

Get in touch

Feedback from our readers is always welcome.

Contact the author: You can visit `https://quantum.allmycode.com`, where Barry Burd (the author) posts additional information as needed. You can even send him an email at `quantum@allmycode.com`. Your reply won't come from a bot or a paid assistant. That's a promise!

General feedback: If you have questions about any aspect of this book, email us at `customercare@packtpub.com` and mention the book title in the subject of your message.

Errata: Although we have taken every care to ensure the accuracy of our content, mistakes do happen. If you have found a mistake in this book, we would be grateful if you would report this to us. Please visit `www.packtpub.com/support/errata` and fill in the form.

Piracy: If you come across any illegal copies of our works in any form on the internet, we would be grateful if you would provide us with the location address or website name. Please contact us at `copyright@packt.com` with a link to the material.

If you are interested in becoming an author: If there is a topic that you have expertise in and you are interested in either writing or contributing to a book, please visit `authors.packtpub.com`.

Share Your Thoughts

Once you've read *Quantum Computing Algorithms*, we'd love to hear your thoughts! Scan the QR code below to go straight to the Amazon review page for this book and share your feedback.

`https://packt.link/r/1-804-61737-7`

Your review is important to us and the tech community and will help us make sure we're delivering excellent quality content.

Download a free PDF copy of this book

Thanks for purchasing this book!

Do you like to read on the go but are unable to carry your print books everywhere? Is your eBook purchase not compatible with the device of your choice?

Don't worry, now with every Packt book you get a DRM-free PDF version of that book at no cost.

Read anywhere, any place, on any device. Search, copy, and paste code from your favorite technical books directly into your application.

The perks don't stop there, you can get exclusive access to discounts, newsletters, and great free content in your inbox daily

Follow these simple steps to get the benefits:

1. Scan the QR code or visit the link below

https://packt.link/free-ebook/9781804617373

2. Submit your proof of purchase

3. That's it! We'll send your free PDF and other benefits to your email directly

Introduction to Quantum Computing

In this brief introduction to quantum computing, we'll cover the following topics:

- What is quantum computing?
- Baby steps toward quantum computing
- Programming a quantum computer
- The future of quantum computing

What is quantum computing?

Quantum reality is very strange.

Imagine a spinning wheel that's turning neither clockwise nor counterclockwise. Just by looking at the wheel, you set up a relationship between yourself and the wheel, and this makes the wheel turn in one direction or another.

Imagine two friends who travel to opposite ends of the Milky Way galaxy. Whatever one randomly decides to say upon landing on a distant planet, the other feels compelled to say too.

That's the world of quantum mechanics. It's what makes quantum computing so fascinating.

Here's how we distinguish quantum computing from classical computing:

- **Classical computing**: A computational model in which the fundamental unit for calculation is a binary bit. Each bit's value is either 0 or 1.

  Every laptop, server, workstation, and smartphone is a kind of classical computer. Even the Frontier supercomputer with 600,000 cores in Oak Ridge, Tennessee is a classical computer. Almost all classical computers in use today are based on **von Neumann's architecture**. With this design, the computer stores sets of instructions (commonly known as **programs**). A processing unit repeatedly fetches the next instruction and then executes it. (What counts as the "next" instruction may depend on the previous instruction's outcome.)

  Alternatives to von Neumann's architecture include Turing machines, which serve mainly as theoretical models, and neural nets, which we will describe briefly in *Chapter 10*.

- **Quantum computing**: A computational model in which the fundamental unit for calculation is a **quantum bit**, commonly known as a **qubit**. A qubit's values can be 0, 1, or values that can roughly be considered "in between" 0 and 1.

  Qubits behave according to the rules of quantum mechanics. Among other things, this means that the previous paragraph's wording ("in between 0 and 1") is a convenient but gross simplification. No matter how you combine a "$\frac{1}{2}$ qubit" with another "$\frac{1}{2}$ qubit," you don't naturally get "1 qubit." The arithmetic that's embodied in a collection of qubits bears little resemblance to our common, everyday ideas about the numbers between 0 and 1.

Baby steps toward quantum computing

The idea for quantum computing came in 1981 with presentations by Paul Benioff and Richard Feynman at the First Conference on the Physics of Computation. Fast forward to 1998, when the world's first quantum computer had only two qubits.

> **Tip**
> For more information about the First Conference, the two-qubit computer, and other topics in this *Introduction*, refer to this chapter's *Further reading* section. You wouldn't buy a laptop whose chip could process only two bits. In the same way, you wouldn't expect a two-qubit quantum computer to solve your puzzling mathematical problems.

By 2006, the world had 12-qubit quantum computers. And by 2017, we had 50-qubit computers. The number of qubits in most advanced quantum computers of the early 2020s is in the low-to-mid hundreds. Compare this with a typical laptop's memory, which stores about 64 billion bits.

Of course, the answer you get when you ask for a count of qubits depends a lot on who you ask. Reports on trustworthy websites sometimes contradict one another. The answer also depends on what you mean when you ask the question. Here are some facts:

- Some specialized quantum computers (described briefly in *Chapter 10*) have 5,000 qubits.
- Today's quantum computers have extra qubits for error checking and error correction, so counting qubits isn't a straightforward task.

  Qubits are extremely fragile, so it's unwise to draw conclusions based on a single qubit's behavior. Quantum theory makes this abundantly clear. To compensate for this inconvenience, physicists and engineers combine the outcomes from several **physical qubits** to come up with an aggregate answer that they call a **logical qubit**. It's like tallying the votes of several physical qubits to come up with one logical qubit representative. But, of course, the "voting" procedure is quite complicated.

  This book is about the logic underlying quantum computing algorithms, not the engineering issues involved in implementing these algorithms on physical machines. So, throughout this book, the word *qubit* refers to a logical qubit. On some quantum computers, what we call a *qubit* might be realized using 10, 100, or even 1,000 physical qubits. A machine that's touted as having 127 qubits may actually need thousands of physical qubits to do its work. As it is with any noisy system, the path to reliability requires some redundancy.

- Some tasks require relatively small numbers of qubits. As I write this chapter in 2023, companies around the world are preparing to make a form of quantum-based cryptography available commercially. The machines that implement this technology have fewer than 100 qubits. (See *Chapter 5* to find out why, in theory, just one qubit at a time will suffice.)

Most quantum algorithms aren't useful unless you have thousands of logical qubits. Millions or billions of logical qubits are even better. So, for now, many of the algorithms described in this book are theoretical, not practical.

But don't underestimate the importance of theoretical results. In 1843, William Hamilton defined a strange number system called the *quaternions*. Today, quaternions help us create 3D computer games. And in 1917, Johann Radon defined his *Radon transform*, which has become an important tool in medical imaging.

Programming a quantum computer

Quantum computers don't run independently. They receive input from classical computers and provide output to classical computers. In a sense, there's no such thing as completely independent quantum computing. All quantum computing is part of a larger technology called **hybrid computing**.

When you work with quantum computers, you write code that runs on a classical computer. Based on your code, the classical computer feeds instructions to the quantum computer. There are many programming platforms designed specifically for quantum computing. They include **Q#** from Microsoft, **Cirq** from Google, **OpenQASM** from IBM, **Ocean** from D-Wave, and **PennyLane**, which is maintained by Xanadu.

In this book, we program using **Qiskit** – an open source software development kit. Qiskit (pronounced *KISS-kit*) is part of IBM's family of quantum computing initiatives. Using Qiskit, you can run code for quantum computers on many different devices. Some of these devices are real quantum computers living in the cloud. Other devices are classical computers that, for small-scale programs, mimic the behavior of real quantum computers.

The future of quantum computing

As far as we know, we'll never trade in all our classical computers for quantum computing models. Quantum computers aren't good for performing the mundane tasks that we assign to most computers today. You wouldn't want to program a simple spreadsheet on a quantum computer, even if you could find a way to do it.

But to solve certain kinds of problems, a quantum computer with sufficiently many qubits will leave classical computers in the dust. *Chapter 9* shows you how sufficiently powerful quantum computers will be able to factor 2,048-bit numbers. According to some estimates, a factoring problem that would take classical computers 300 trillion years to solve will require only 10 seconds of a quantum computer's time. If we can achieve an advantage of this kind using a real quantum computer, we call it **quantum supremacy**.

In 2019, a team at Google claimed to have demonstrated quantum supremacy. Its 53-qubit quantum computer took about 3 minutes to solve a problem that supposedly would have required 47 classical computing years. But in 2022, researchers in China showed how a classical computer could solve the same problem in about a minute. So much for Google's claim!

In 2020, a group at the University of Science and Technology of China staked its own quantum supremacy claim. Using 76 qubits, it generated a particular set of random numbers that it said would take classical computers nearly 1 billion years to generate. As of mid-2023, the group's claim has not been debunked.

With or without any kind of supremacy, the field of quantum computing presents exciting possibilities. Engineers, physicists, and computer scientists are hard at work finding new ways to leverage the technology's capabilities. Market research firms predict that the industry's growth rate will be at least 30 percent per year. Start-ups are preparing for applications in healthcare and finance. The last time I visited LinkedIn, I found more than 2,500 quantum computing job listings in the United States, with over 6,500 listings worldwide.

So, it's time to learn more about quantum computing. If I were you, I'd move on to *Chapter 1*.

Further reading

1. Benioff, Paul A. (April 1, 1982). "*Quantum mechanical Hamiltonian models of discrete processes that erase their own histories: Application to Turing machines*". International Journal of Theoretical Physics. 21 (3): 177–201. Bibcode:1982IJTP...21..177B. doi:10.1007/BF01857725. ISSN 1572-9575. S2CID 122151269.

2. Chuang, Isaac L.; Gershenfeld, Neil; Kubinec, Mark (April 13, 1998). "*Experimental Implementation of Fast Quantum Searching*". Physical Review Letters. 80 (15): 3408–3411. Bibcode:1998PhRvL..80.3408C. doi:10.1103/PhysRevLett.80.3408. S2CID 13891055.

3. Feynman, R.P.(1982). *Simulating physics with computers.* Int J Theor Phys 21, 467–488 (1982). `https://doi.org/10.1007/BF02650179`

4. *Quantum computers could crack today's encrypted messages. That's a problem*: `https://www.cnet.com/tech/computing/quantum-computers-could-crack-todays-encrypted-messages-thats-a-problem/`

5. Arute, F., Arya, K., Babbush, R. et al. *Quantum supremacy using a programmable superconducting processor.* Nature 574, 505–510 (2019). `https://doi.org/10.1038/s41586-019-1666-5`

6. Pan, Feng; Chen, Keyang; Zhang, Pan (2022). *"Solving the Sampling Problem of the Sycamore Quantum Circuits"*. Physical Review Letters. 129 (9): 090502. arXiv:2111.03011. Bibcode:2022PhRvL.129i0502P. doi:10.1103/PhysRevLett.129.090502. PMID 36083655. S2CID 251755796.

7. Zhong, Han-Sen; Wang, Hui; Deng, Yu-Hao; Chen, Ming-Cheng; Peng, Li-Chao; Luo, Yi-Han; Qin, Jian; Wu, Dian; Ding, Xing; Hu, Yi; Hu, Peng (2020-12-03). *"Quantum computational advantage using photons"*. Science. 370 (6523): 1460–1463. arXiv:2012.01625. Bibcode:2020Sci...370.1460Z. doi:10.1126/science.abe8770. ISSN 0036-8075. PMID 33273064. S2CID 227254333.

8. *Statistics on Global Quantum Computing Market Size & Share to Surpass USD 5274.9 Mn by 2030, Exhibit a CAGR of 31.21% | Quantum Computing Industry Trends, Value, Analysis & Forecast Report by Zion Market Research*: https://www.prnewswire.com/news-releases/statistics-on-global-quantum-computing-market-size--share-to-surpass-usd-5274-9-mn-by-2030--exhibit-a-cagr-of-31-21--quantum-computing-industry-trends-value-analysis--forecast-report-by-zion-market-research-301724817.html

Part 1
Nuts and Bolts

This part lays the foundation for the exploration of quantum computing. You'll learn about two weird features of quantum reality: superposition and entanglement. You'll reinforce what you learn by writing code for quantum computers and performing calculations to back up your intuitions.

This part has the following chapters:

- *Chapter 1, New Ways to Think about Bits*
- *Chapter 2, What Is a Qubit?*
- *Chapter 3, Math for Qubits and Quantum Gates*
- *Chapter 4, Qubit Conspiracy Theories*

1
New Ways to Think about Bits

Classical computing is everywhere. The phone in your pocket, the laptop on your desk, and the world's fastest supercomputers use classical computing hardware. A classical computer codes everything in bits, and bits are quite simple. The rules for dealing with bits have been known since the mid-1800s. (George Boole wrote his ground-breaking work on logic in 1854). The computer hardware that manipulates bits has developed steadily since the mid-1940s.

Before you read about quantum computing, you need to have explored certain mathematical concepts. With that in mind, this chapter shows you how math applies to ordinary, classical bits.

If you've done any coding, you're already familiar with the **and**, **or**, and **not** operators, but you may not know how to represent these operators using matrices. This chapter explores the connection between bit operations and matrices. After a tour through the math, we walk through the creation of matrices using Jupyter notebooks with Python.

In this chapter, we'll cover the following main topics:

- Bits and logic gates
- Working with matrices
- Matrix representation of bits and gates
- Jupyter notebooks
- Matrices in Python

Technical requirements

Every chapter in this book requires that you have access to a web browser. You can download each chapter's Python code from the following GitHub link: `https://github.com/PacktPublishing/Quantum-Computing-Algorithms`.

Bits and logic gates

A classical computer codes information as sequences of **bits**, each bit having one of two values – 0 or 1. Of course, if you looked inside the circuitry, you wouldn't see things shaped like ovals for 0s and vertical lines for 1s. Instead, you might observe electrical voltages. Any voltage below 0.8 volts might stand for a 0 bit, and any voltage over 2.0 volts might stand for 1 bit. A voltage between 0.8 and 2.0 would be treated as an error condition.

As bits flow from one part of the computer to another, they pass through the circuitry's gates. A **gate** is a piece of hardware (often made of transistors) that operates on bits. Each kind of gate is named after the operation that it performs.

Let's see a few examples:

- A **NOT gate** performs a **NOT operation**. The gate outputs **1** *if and only if* the gate's input is not **1**:

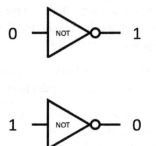

Figure 1.1 – Behavior of the NOT gate

- An **AND gate** performs the **AND operation** with two inputs. The gate outputs **1** if and only if the first input is **1** and the second input is **1**:

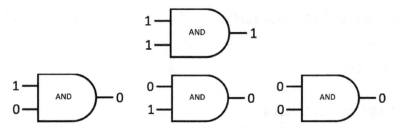

Figure 1.2 – Behavior of the AND gate

- An **OR gate** performs the **OR operation** with two inputs. The gate outputs **1** if and only if the first input or the second input, or both inputs are **1**:

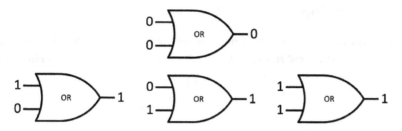

Figure 1.3 – Behavior of the OR gate

Notice the very last OR gate example. In English, a sentence such as "*It's raining, or it's cold*" usually means "*Either it's raining, or it's cold, but it's not both raining and cold.*" With classical gates, the word *OR* means "*Either the topmost bit is 1 or the bottommost bit is 1, or both bits are 1.*"

Taken together, the NOT, AND, and OR gates form a **universal set**. This means you can perform any logical operation using only NOT, AND, and OR gates.

For example, *Figure 1.4* shows the addition of one-bit binary numbers:

$$
\begin{array}{cccc}
0 & 0 & 1 & 1 \\
+0 & +1 & +0 & +1 \\
\hline
0\,0 & 0\,1 & 0\,1 & 1\,0 \\
\end{array}
$$

Figure 1.4 – Binary addition

Figure 1.5 shows a circuit that adds one-bit binary numbers:

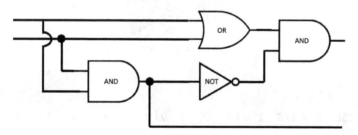

Figure 1.5 – Adding binary numbers using logic gates

Full disclosure

As human beings, we tend to reason using NOT, AND, and OR operators, but using NOT, AND, and OR gates isn't always the most efficient way to build computer circuitry. Many circuits make heavy use of NAND gates (think *NOT BOTH* gates) since the NAND gate alone forms a universal set. Another kind of gate, called a NOR gate, also forms a universal set (for a NOR gate, think *NEITHER ... NOR*).

Binary representation

Our Arabic number system is biased by the human mind's affinity with the number 10. For example, to represent the value *four-hundred-sixty-three*, we write *463*. *Figure 1.6* shows you how it works:

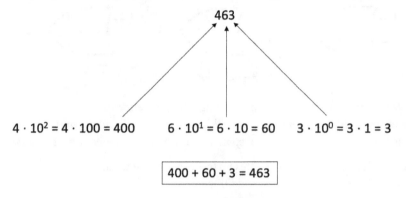

Figure 1.6 – Decimal number representation

This way of representing numbers is called the **decimal (base 10)** system because the prefix *dec* comes from the Greek word for *ten*. The numbers 0, 1, 2, 3, 4, 5, 6, 7,8, and 9 are called digits.

If we ever meet intelligent beings from another planet, they probably won't use the decimal system. They may, in fact, use the **binary (base 2)** system. In the binary system, we represent *four-hundred-sixty-three* with the string *111001111*, as shown in *Figure 1.7*:

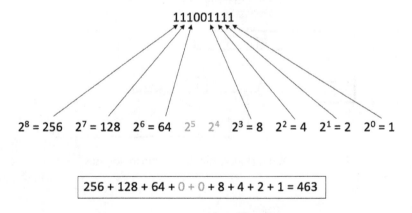

Figure 1.7 – Binary number representation

This way of representing numbers is called the **binary (base 2) system** because the prefix *bi* comes from the Latin word for *two*. The numbers 0 and 1 are called **binary digits**, also known as **bits**.

So, when you see the characters *100*, what do they mean? Are they the binary representation of the number *four*, or the decimal representation of the number *one hundred*? If there's any doubt, I use a subscript to indicate the number's base. For example, 100_2 is the binary representation of *four*, and 100_{10} is the decimal representation of *one hundred*.

The table in *Figure 1.8* shows 0 to 15 in their decimal and binary representations. It's handy if you can easily recognize these four-bit binary representations, but don't bother memorizing the numbers in the table. Instead, familiarize yourself with these numbers by looking for patterns. Take some time to do it. It's worth the effort.

Decimal	Binary
0	0000
1	0001
2	0010
3	0011
4	0100
5	0101
6	0110
7	0111
8	1000
9	1001
10	1010
11	1011
12	1100
13	1101
14	1110
15	1111

Figure 1.8 – Decimal versus binary

The next few sections describe a way to represent bits and logic gates in a particular numeric form.

Working with matrices

A **matrix** is a rectangle of numbers. For example:

$$\begin{pmatrix} 2 & -3 & 0 \\ 1 & 5 & 19 \end{pmatrix}$$

The matrix shown previously has two rows and three columns, so we call it a **2×3 matrix** (pronounced as *two-by-three matrix*).

> **Tip**
> The plural of *matrix* is *matrices* (pronounced as *MAY-trih-sees*). To sound like a pro, never say *matrixes* or *matricee*.

It's common to use an uppercase to represent a matrix. You refer to the entries in a matrix using the entries' row numbers and column numbers. Some authors start with row number 1 and column number 1, but, for our purposes, it's better to start with row number 0 and column number 0.

$$A = \begin{pmatrix} a_{00} & a_{01} & a_{02} \\ a_{10} & a_{11} & a_{12} \end{pmatrix} = \begin{pmatrix} 2 & -3 & 0 \\ 1 & 5 & 19 \end{pmatrix}$$

When you talk about matrices, you need a name that refers to a single number – a number that isn't inside of a matrix. For example, if you write the number 12 with no parentheses around it, you're referring to one of these single numbers. A number of this kind is called a **scalar**.

So, what can you do with a scalar? One thing you can do is multiply a scalar by a matrix. To do so, you multiply the scalar by each of the matrix's entries. For example:

$$4 \begin{pmatrix} 2 & -3 & 0 \\ 1 & 5 & 19 \end{pmatrix} = \begin{pmatrix} 8 & -12 & 0 \\ 4 & 20 & 56 \end{pmatrix}$$

One way to work with bits and logic gates is to represent them as matrices. From this point of view, combining gates is the same as finding products of matrices. With that in mind, the rest of this section covers various kinds of matrix products.

Vectors

When a matrix has only one row or only one column, we give it a special name. We call it a **vector**. Let's see the following examples:

- This matrix is a **row vector**:

$$\begin{pmatrix} 8 & 3 & -2 & 1.5 \end{pmatrix}$$

- And this matrix is a **column vector**:

$$\begin{pmatrix} 4.3 \\ 4.91 \\ -0.55 \\ 2 \\ 1.7 \end{pmatrix}$$

It's common to refer to a vector's entries using the entries' row numbers or column numbers. For our purposes, it's best to start with row number 0 or column number 0:

$$\begin{pmatrix} a_0 \\ a_1 \\ a_2 \\ a_3 \\ a_4 \end{pmatrix} = \begin{pmatrix} 4.3 \\ 4.91 \\ -0.55 \\ 2 \\ 1.7 \end{pmatrix}$$

In classical computing, we can do the following:

- Represent a bit with a column vector

- Represent a collection of bits with a column vector

- Represent an operation on bits with a matrix

The same is true for quantum computing. You can read more about this in the section entitled *Matrix representation of bits and gates*.

When you have two vectors of the same size (one row vector and one column vector), you can combine them by calculating the **dot product**. To do this, you zipper the two vectors together. You multiply the vectors' corresponding entries and then add up the products. For example:

$$\begin{pmatrix} 5 & 1 & 3 \end{pmatrix} \begin{pmatrix} 2 \\ 4 \\ 6 \end{pmatrix} \Rightarrow \begin{pmatrix} 5 \cdot 2 \\ + \\ 1 \cdot 4 \\ + \\ 3 \cdot 6 \end{pmatrix} \Rightarrow \begin{pmatrix} 10 \\ + \\ 4 \\ + \\ 18 \end{pmatrix} \Rightarrow 32$$

> **Important note**
>
> In order to take a dot product, you must start with two vectors of the same size. You can't take the dot product of a four-entry row vector and a six-entry column vector.

Matrix multiplication

You can think of a matrix as a combination of vectors. When you do, you can multiply matrices using repeated dot products. For example, start with the following two matrices:

$$\begin{pmatrix} 1 & 3 & 2 \\ 4 & 0 & -1 \end{pmatrix} \begin{pmatrix} 5 & 10 \\ 2 & -2 \\ -3 & 6 \end{pmatrix}$$

Think of these matrixes as combinations of vectors, as shown here:

$$\begin{pmatrix} \boxed{1 \quad 3 \quad 2} \\ \boxed{4 \quad 0 \quad -1} \end{pmatrix} \left(\boxed{\begin{matrix} 5 \\ 2 \\ -3 \end{matrix}} \boxed{\begin{matrix} 10 \\ -2 \\ 6 \end{matrix}} \right)$$

Combine each row on the left with each column on the right (that is, find all the dot products). Use the dot products to form a new matrix. The new matrix's entries go in positions corresponding to the row number and column numbers in the original two matrices.

For example, the dot product of **Row 0** on the left and **Column 0** on the right gives you the entry in the new matrix's row 0, column 0 position:

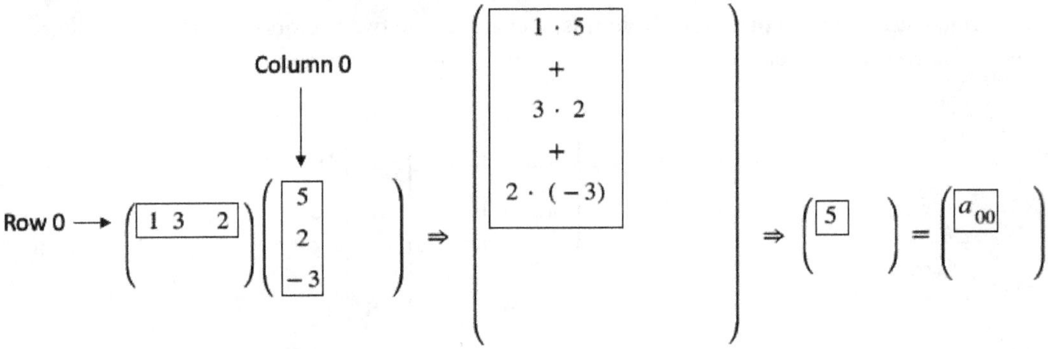

Figure 1.9 – Multiply Row 0 by Column 0

In general, you get the new matrix's a_{ij} entry by taking the dot product of row i on the left with column j on the right:

Column 0
Column 1
Row 0
Row 1

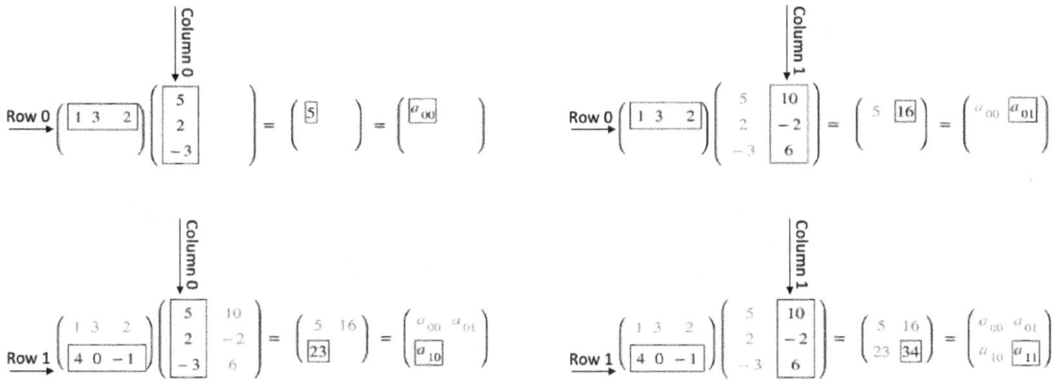

Figure 1.10 – In general, multiply Row i by Column j

> **Important note**
>
> To multiply two matrices, the number of columns in one matrix (the matrix on the left) must equal the number of rows in the other matrix (the matrix on the right). You can multiply a *5x7* matrix by a *7x10* matrix, but you can't take the product of a *5x7* matrix and a *6x10* matrix because 7 doesn't equal 6.

At this point, I must emphasize how important it is to distinguish left from right. Remember that when you multiply matrices, you combine each row on the left with each column on the right. You don't do it the other way around.

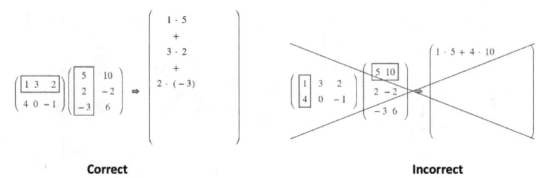

Correct
Incorrect

Figure 1.11 – Multiply the left matrix's rows by the right matrix's columns

Also, matrix multiplication isn't commutative. In other words, order matters. Look at these two products:

$$\begin{pmatrix} 1 & 0 \\ 2 & 1 \end{pmatrix}\begin{pmatrix} 3 & 3 \\ 1 & 2 \end{pmatrix} = \begin{pmatrix} 3 & 3 \\ 7 & 8 \end{pmatrix} \qquad \begin{pmatrix} 3 & 3 \\ 1 & 2 \end{pmatrix}\begin{pmatrix} 1 & 0 \\ 2 & 1 \end{pmatrix} = \begin{pmatrix} 9 & 3 \\ 5 & 2 \end{pmatrix}$$

Swapping the left and right matrices changes the final answer, and there's no simple rule for turning one final answer into the other.

> **Important note**
>
> There are several ways to multiply matrices. The way that's shown in this section is simply called **multiplying** or **finding the product**. For emphasis, I sometimes refer to this kind of multiplication as **ordinary matrix multiplication** or **finding the dot product** of two matrices. My terminology isn't standard. Most authors refer only to the *dot product* of two vectors and the *product* of two matrices.

The tensor product

For a moment, let's set ordinary matrix multiplication aside. Another way to multiply matrices goes by the name **tensor product**. To find the tensor product of two matrices, you perform a scalar multiplication of each entry in the left matrix with the entire matrix on the right. Here are some examples:

- To find the tensor product $\begin{pmatrix} 3 \\ 2 \end{pmatrix} \otimes \begin{pmatrix} 5 \\ 4 \end{pmatrix}$, multiply 3 by $\begin{pmatrix} 5 \\ 4 \end{pmatrix}$ and multiply 2 by $\begin{pmatrix} 5 \\ 4 \end{pmatrix}$:

$$\begin{pmatrix} 3 \\ 2 \end{pmatrix} \otimes \begin{pmatrix} 5 \\ 4 \end{pmatrix} = \begin{pmatrix} 3\begin{pmatrix} 5 \\ 4 \end{pmatrix} \\ 2\begin{pmatrix} 5 \\ 4 \end{pmatrix} \end{pmatrix} = \begin{pmatrix} 3 \cdot 5 \\ 3 \cdot 4 \\ 2 \cdot 5 \\ 2 \cdot 4 \end{pmatrix} = \begin{pmatrix} 15 \\ 12 \\ 10 \\ 8 \end{pmatrix}$$

- To find the tensor product $\begin{pmatrix} 5 & 1 \\ 2 & -2 \\ -3 & 6 \end{pmatrix} \otimes \begin{pmatrix} 3 & 0 \\ 4 & 10 \end{pmatrix}$, multiply each entry in the left matrix by $\begin{pmatrix} 3 & 0 \\ 4 & 10 \end{pmatrix}$:

$$\begin{pmatrix} 5 & 1 \\ 2 & -2 \\ -3 & 6 \end{pmatrix} \otimes \begin{pmatrix} 3 & 0 \\ 4 & 10 \end{pmatrix} = \begin{pmatrix} 5\begin{pmatrix} 3 & 0 \\ 4 & 10 \end{pmatrix} & 1\begin{pmatrix} 3 & 0 \\ 4 & 10 \end{pmatrix} \\ 2\begin{pmatrix} 3 & 0 \\ 4 & 10 \end{pmatrix} & -2\begin{pmatrix} 3 & 0 \\ 4 & 10 \end{pmatrix} \\ -3\begin{pmatrix} 3 & 0 \\ 4 & 10 \end{pmatrix} & 6\begin{pmatrix} 3 & 0 \\ 4 & 10 \end{pmatrix} \end{pmatrix} = \begin{pmatrix} 15 & 0 & 3 & 0 \\ 20 & 50 & 4 & 10 \\ 6 & 0 & -6 & 0 \\ 8 & 20 & -8 & -20 \\ -9 & 0 & 18 & 0 \\ -12 & -30 & 24 & 60 \end{pmatrix}$$

Notice that the matrices in a tensor product don't have any size requirements. You can find the tensor product of a *3x5* matrix with a *10x4* matrix. When you do, you get a *30x20* matrix.

Important note

Remember that when you find a tensor product, you start by spreading the leftmost matrix's entries and keeping the rightmost matrix's entries grouped. You don't do it the other way around.

For example:

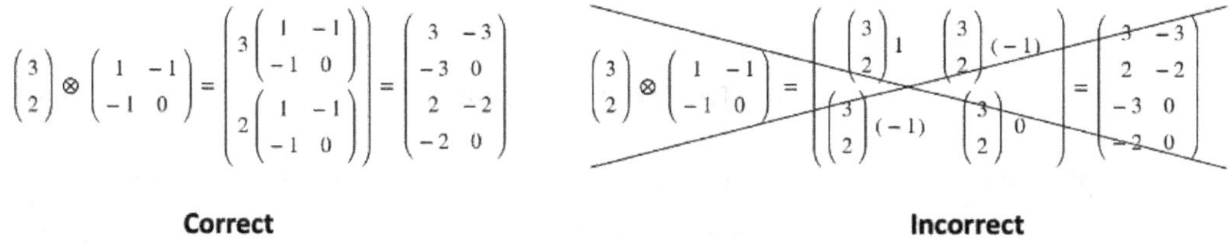

$$\begin{pmatrix} 3 \\ 2 \end{pmatrix} \otimes \begin{pmatrix} 1 & -1 \\ -1 & 0 \end{pmatrix} = \begin{pmatrix} 3\begin{pmatrix} 1 & -1 \\ -1 & 0 \end{pmatrix} \\ 2\begin{pmatrix} 1 & -1 \\ -1 & 0 \end{pmatrix} \end{pmatrix} = \begin{pmatrix} 3 & -3 \\ -3 & 0 \\ 2 & -2 \\ -2 & 0 \end{pmatrix}$$

$$\begin{pmatrix} 3 \\ 2 \end{pmatrix} \otimes \begin{pmatrix} 1 & -1 \\ -1 & 0 \end{pmatrix} = \begin{pmatrix} \begin{pmatrix} 3 \\ 2 \end{pmatrix}1 & \begin{pmatrix} 3 \\ 2 \end{pmatrix}(-1) \\ \begin{pmatrix} 3 \\ 2 \end{pmatrix}(-1) & \begin{pmatrix} 3 \\ 2 \end{pmatrix}0 \end{pmatrix} = \begin{pmatrix} 3 & -3 \\ 2 & -2 \\ -3 & 0 \\ -2 & 0 \end{pmatrix}$$

Correct   **Incorrect**

Figure 1.12 – Multiply each entry in the left matrix by the entire right matrix

Like ordinary matrix multiplication, taking a tensor product isn't commutative. For example:

$$\begin{pmatrix} 3 \\ 2 \end{pmatrix} \otimes \begin{pmatrix} 1 & -1 \\ -1 & 0 \end{pmatrix} = \begin{pmatrix} 3\begin{pmatrix} 1 & -1 \\ -1 & 0 \end{pmatrix} \\ 2\begin{pmatrix} 1 & -1 \\ -1 & 0 \end{pmatrix} \end{pmatrix} = \begin{pmatrix} 3 & -3 \\ -3 & 0 \\ 2 & -2 \\ -2 & 0 \end{pmatrix}$$

$$\begin{pmatrix} 1 & -1 \\ -1 & 0 \end{pmatrix} \otimes \begin{pmatrix} 3 \\ 2 \end{pmatrix} = \begin{pmatrix} 1\begin{pmatrix} 3 \\ 2 \end{pmatrix} & 1\begin{pmatrix} 3 \\ 2 \end{pmatrix} \\ -1\begin{pmatrix} 3 \\ 2 \end{pmatrix} & 0\begin{pmatrix} 3 \\ 2 \end{pmatrix} \end{pmatrix} = \begin{pmatrix} 3 & 3 \\ 2 & 2 \\ -3 & 0 \\ -2 & 0 \end{pmatrix}$$

The entries in one answer are rearranged in the other.

Combining gates and bits

What do matrix operations have to do with computer logic? When you create a circuit, there are two ways to combine gates:

- When you combine gates **in series**, you apply the gates one after another to the same stream of bits. To do the math, you use ordinary matrix multiplication.

Figure 1.13 – Two operations in series

- When you combine gates **in parallel**, you apply the gates side by side to different streams of bits. To do the math, you use a tensor product.

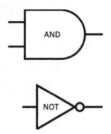

Figure 1.14 – Two operations in parallel

You can also apply a gate to its input bits. When you do, you use ordinary matrix multiplication.

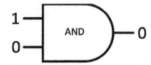

Figure 1.15 – Sending bits through a gate

You can read more about this in the next section.

Matrix representation of bits and gates

For an inkling of the way matrices work in computer logic, we introduce two new ways to represent bits:

- In **Dirac notation**, the zero bit is $|0\rangle$, and the one bit is $|1\rangle$.

  The $|\ \rangle$ combination of characters is called a **ket**.
- In **vector notation**, the zero bit is $\begin{pmatrix} 1 \\ 0 \end{pmatrix}$, and the one bit is $\begin{pmatrix} 0 \\ 1 \end{pmatrix}$.

These new ways to represent bits may seem cumbersome and redundant, but they're really very helpful. If you like, think of the numbers in a vector as amounts ranging from zero to one. A vector's top entry is an *amount of zero-ness* and the vector's bottom entry is *an amount of one-ness*.

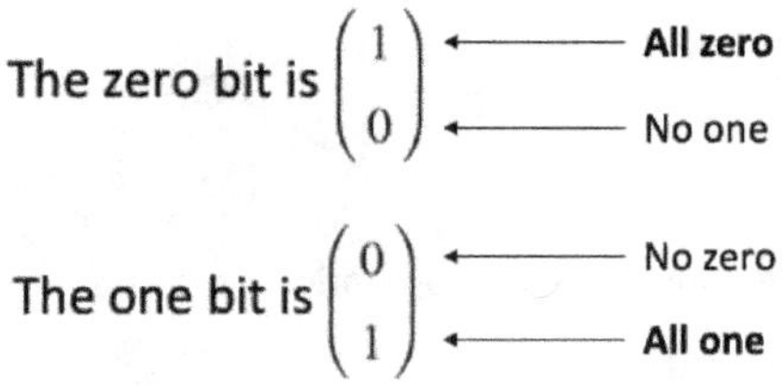

$$\text{The zero bit is } \begin{pmatrix} 1 \\ 0 \end{pmatrix} \begin{array}{l} \longleftarrow \textbf{All zero} \\ \longleftarrow \text{No one} \end{array}$$

$$\text{The one bit is } \begin{pmatrix} 0 \\ 1 \end{pmatrix} \begin{array}{l} \longleftarrow \text{No zero} \\ \longleftarrow \textbf{All one} \end{array}$$

Figure 1.16 – The correspondence between vector notation and Dirac notation

This business about *all zero-ness* and *all one-ness* will make more sense when you read about qubits in the next chapter.

> **Disclaimer**
>
> Most authors reserve kets and vectors for qubits (quantum bits). For classical bits, they use the unadorned symbols *0* and *1*. I'll follow that rule in most chapters, but in this chapter, the Dirac and vector notations for bits are handy.

We represent a bit with a vector and represent an operator with a matrix. For example, the NOT operator is the matrix $\begin{pmatrix} 0 & 1 \\ 1 & 0 \end{pmatrix}$:

$$NOT \; |0\rangle = \begin{pmatrix} 0 & 1 \\ 1 & 0 \end{pmatrix}\begin{pmatrix} 1 \\ 0 \end{pmatrix} = \begin{pmatrix} 0 \\ 1 \end{pmatrix} = |1\rangle$$

$$NOT \; |1\rangle = \begin{pmatrix} 0 & 1 \\ 1 & 0 \end{pmatrix}\begin{pmatrix} 0 \\ 1 \end{pmatrix} = \begin{pmatrix} 1 \\ 0 \end{pmatrix} = |0\rangle$$

To describe the AND and OR operators, we need ways of representing combinations of two bits, three bits, or any other number of bits:

- In Dirac notation, there are four different combinations consisting of two bits. We can write the four combinations this way:

$$|0\rangle|0\rangle \qquad |0\rangle|1\rangle \qquad |1\rangle|0\rangle \qquad |1\rangle|1\rangle$$

Or we can abbreviate the four combinations this way:

$$|00\rangle \qquad |01\rangle \qquad |10\rangle \qquad |11\rangle$$

If you interpret these as binary numbers, you have 0, 1, 2, and 3. In the same way, the three-bit ket $|100\rangle$ represents the binary number 4, $|101\rangle$ represents binary 5, and so on.

- In vector notation, you represent pairs of bits by forming a tensor product:

$$|00\rangle \text{ is } \begin{pmatrix} 1 \\ 0 \end{pmatrix} \otimes \begin{pmatrix} 1 \\ 0 \end{pmatrix} = \begin{pmatrix} 1 \\ 0 \\ 0 \\ 0 \end{pmatrix}$$

$$|01\rangle \text{ is } \begin{pmatrix} 1 \\ 0 \end{pmatrix} \otimes \begin{pmatrix} 0 \\ 1 \end{pmatrix} = \begin{pmatrix} 0 \\ 1 \\ 0 \\ 0 \end{pmatrix}$$

$$|10\rangle \text{ is } \begin{pmatrix} 0 \\ 1 \end{pmatrix} \otimes \begin{pmatrix} 1 \\ 0 \end{pmatrix} = \begin{pmatrix} 0 \\ 0 \\ 1 \\ 0 \end{pmatrix}$$

$$|11\rangle \text{ is } \begin{pmatrix} 0 \\ 1 \end{pmatrix} \otimes \begin{pmatrix} 0 \\ 1 \end{pmatrix} = \begin{pmatrix} 0 \\ 0 \\ 0 \\ 1 \end{pmatrix}$$

You do the same thing when you have more than two bits:

$$|110\rangle \quad = \begin{pmatrix} 0 \\ 1 \end{pmatrix} \otimes \begin{pmatrix} 0 \\ 1 \end{pmatrix} \otimes \begin{pmatrix} 1 \\ 0 \end{pmatrix} = \begin{pmatrix} 0 \\ 1 \end{pmatrix} \otimes \begin{pmatrix} 0 \\ 0 \\ 1 \\ 0 \end{pmatrix} = \begin{pmatrix} 0 \\ 0 \\ 0 \\ 0 \\ 0 \\ 0 \\ 1 \\ 0 \end{pmatrix}$$

Once again, it's helpful to think of bit combinations as binary number representations. Here's an example:

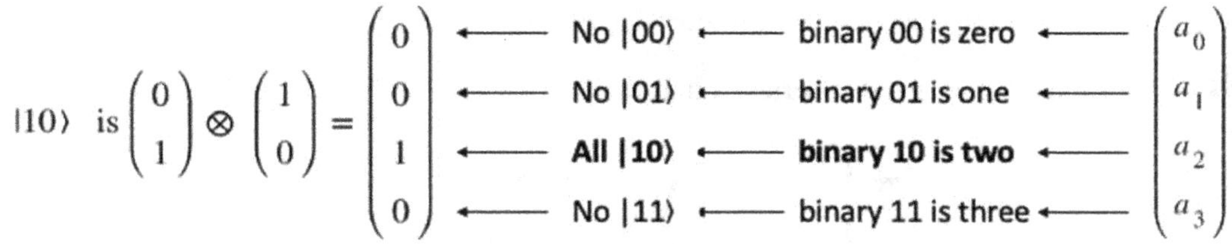

$$|10\rangle \text{ is } \begin{pmatrix} 0 \\ 1 \end{pmatrix} \otimes \begin{pmatrix} 1 \\ 0 \end{pmatrix} = \begin{pmatrix} 0 \\ 0 \\ 1 \\ 0 \end{pmatrix}$$

- 0 ←— No $|00\rangle$ ←— binary 00 is zero ←— a_0
- 0 ←— No $|01\rangle$ ←— binary 01 is one ←— a_1
- 1 ←— **All $|10\rangle$** ←— **binary 10 is two** ←— a_2
- 0 ←— No $|11\rangle$ ←— binary 11 is three ←— a_3

Figure 1.17 – Example of bit combinations as binary number representation

With this in mind, the AND operator's matrix representation is $\begin{pmatrix} 1 & 1 & 1 & 0 \\ 0 & 0 & 0 & 1 \end{pmatrix}$. For example:

$$|1\rangle \ AND \ |0\rangle = |0\rangle \quad \text{is the same as} \quad AND \ |10\rangle = \begin{pmatrix} 1 & 1 & 1 & 0 \\ 0 & 0 & 0 & 1 \end{pmatrix} \begin{pmatrix} 0 \\ 0 \\ 1 \\ 0 \end{pmatrix} = \begin{pmatrix} 1 \\ 0 \end{pmatrix} = |0\rangle$$

$$|1\rangle \ AND \ |1\rangle = |1\rangle \quad \text{is the same as} \quad AND \ |11\rangle = \begin{pmatrix} 1 & 1 & 1 & 0 \\ 0 & 0 & 0 & 1 \end{pmatrix} \begin{pmatrix} 0 \\ 0 \\ 0 \\ 1 \end{pmatrix} = \begin{pmatrix} 0 \\ 1 \end{pmatrix} = |1\rangle$$

The OR operator's matrix representation is $\begin{pmatrix} 1 & 0 & 0 & 0 \\ 0 & 1 & 1 & 1 \end{pmatrix}$. For example:

$$|0\rangle \ OR \ |0\rangle = |0\rangle \quad \text{is the same as} \quad OR \ |00\rangle = \begin{pmatrix} 1 & 0 & 0 & 0 \\ 0 & 1 & 1 & 1 \end{pmatrix} \begin{pmatrix} 1 \\ 0 \\ 0 \\ 0 \end{pmatrix} = \begin{pmatrix} 1 \\ 0 \end{pmatrix} = |0\rangle$$

$$|1\rangle \ OR \ |0\rangle = |1\rangle \quad \text{is the same as} \quad OR \ |10\rangle = \begin{pmatrix} 1 & 0 & 0 & 0 \\ 0 & 1 & 1 & 1 \end{pmatrix} \begin{pmatrix} 0 \\ 0 \\ 1 \\ 0 \end{pmatrix} = \begin{pmatrix} 0 \\ 1 \end{pmatrix} = |1\rangle$$

In this section, we used matrices to represent individual gates. In the next section, we will represent combinations of gates using matrices.

Matrix operations and computer logic

Now, we can revisit some rules that we introduced in the previous section:

- When you combine gates **in series**, you use ordinary matrix multiplication:

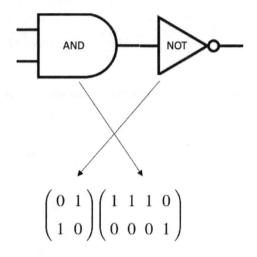

$$\begin{pmatrix} 0 & 1 \\ 1 & 0 \end{pmatrix} \begin{pmatrix} 1 & 1 & 1 & 0 \\ 0 & 0 & 0 & 1 \end{pmatrix}$$

Figure 1.18 – Two gates in series

- When you combine gates **in parallel**, you use a tensor product.

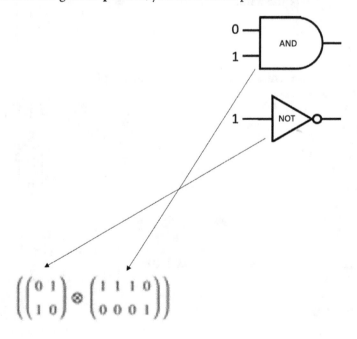

$$\left(\begin{pmatrix} 0 & 1 \\ 1 & 0 \end{pmatrix} \otimes \begin{pmatrix} 1 & 1 & 1 & 0 \\ 0 & 0 & 0 & 1 \end{pmatrix} \right)$$

Figure 1.19 – Two gates in parallel

- When you apply gates to bits, you use ordinary matrix multiplication:

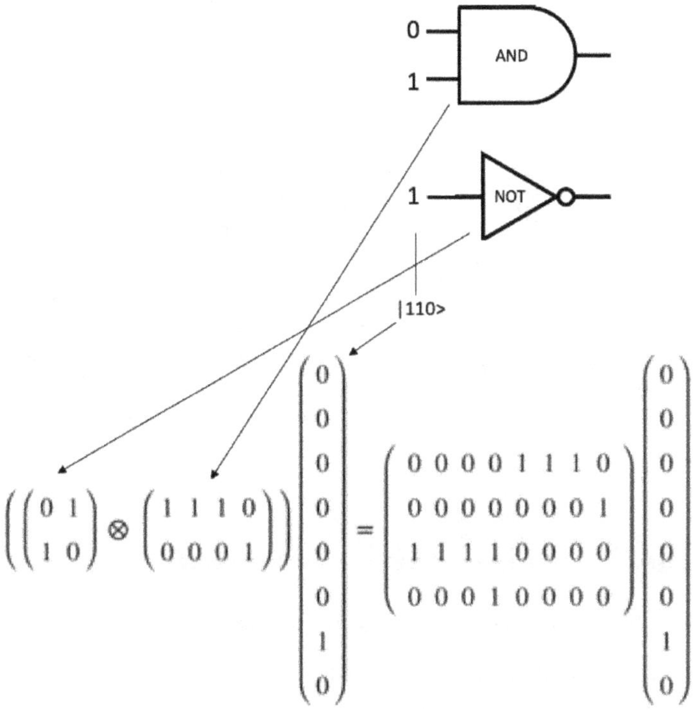

Figure 1.20 – Sending bits through gates

Enough about math! This book is about quantum computing. So, in the next section, we start doing some computing.

Jupyter notebooks

Jupyter notebooks are web-based environments for running code and displaying information about the code. To program quantum computers, we'll create Jupyter notebooks that run Python programs with IBM's Qiskit library functions.

To get started with Qiskit, follow these steps:

1. Visit the quantum computing lab: `https://quantum-computing.ibm.com/lab`.

2. Create an IBM account.

3. Return to the quantum computing lab page: `https://quantum-computing.ibm.com/lab`.

4. In the resulting **Launcher** tab, click on the **Notebook | Python 3 (ipykernel)** buttons:

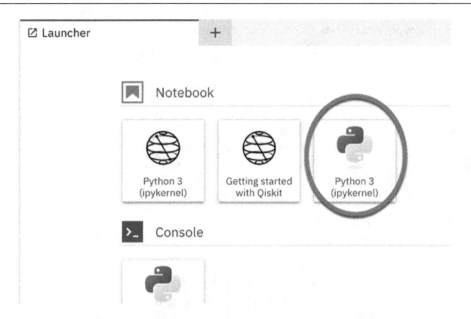

Figure 1.21 – Creating a Qiskit notebook

A new Jupyter notebook appears. Near the top of the notebook, there's an empty text field.

In the next section, you will populate that text field with Python code.

Creating and displaying values

A Jupyter notebook consists of **cells**. In a typical scenario, each cell contains a text field. You can type code in the text field. Then, with your cursor positioned anywhere in the text field, you press *Shift+Enter*. The *Shift+Enter* key combination makes a server run the code. The code's output (if any) appears below the text field. In addition, a new empty cell appears.

In a Jupyter notebook, the order in which cells appear doesn't matter very much. The notebook executes code only when you press *Shift+Enter*. To see this in action, try the following experiments:

1. In the cell, at the top of the notebook, type x = 7, and then press *Shift+Enter*.

    No output appears, but the notebook has a second (empty) cell.

2. In the second cell, type print (x), and then press *Shift+Enter*.

    With the execution of the print call, the output 7 appears below the text field in the cell. Also, a third cell appears.

> **Important note**
>
> Lines such as x = 7 and print(x) are statements in the Python programming language. You don't have to know much Python to follow this book's examples, but if you need some Python programming practice, get a copy of *Learn Python Programming* by Fabrizio Romano and Heinrich Kruger.

3. Place your cursor back in the first cell. Change x = 7 to x = 99, but **don't** press *Shift+Enter*.

4. Place your cursor back in the print(x) cell, and then press *Shift+Enter*.

   With the execution of the print call, the output 7 appears again. Typing x = 99 in the cell above the print cell has no effect because you didn't execute x = 99.

5. Place your cursor back in the first cell. With x = 99 in that cell, press *Shift+Enter*.

   Python has now set x to 99, but the second cell still displays the number 7.

6. Place your cursor back in the print(x) cell, and then press *Shift+Enter*.

   Finally, the number 99 appears in the second cell's output.

7. In the third cell, type x = 1024, and then press *Shift+Enter*.

   Yet another empty cell appears below the third cell.

8. Place your cursor back in the print(x) cell, and then press *Shift+Enter*.

   The output 1024 appears. The notebook doesn't care whether x = 1024 appears before or after print(x) in the notebook. The only thing that matters is the order in which the cells receive your *Shift+Enter* key presses.

Keep this notebook open. You'll add more code to it in the next few sections.

> **Tip**
>
> You can select one or more cells in a Jupyter notebook. To select a single cell, click any part of the cell that's not inside the cell's text field. To select several cells at once, select one of the cells and then *Shift+click* another. With one or more cells selected, pressing *Ctrl+Enter* instructs the notebook to execute *all* of the selected cells.

Which values are (or are not) defined?

Sometimes, when you call the print function, Python doesn't print a value. Instead, Python complains that you made a bad call. To see this, follow the next few steps. Continue from where you left off in the *Creating and displaying values* section's notebook:

1. Right-click anywhere in the body of the notebook. In the resulting context menu, select **Restart Kernel**.

The x variable no longer has a value.

2. Place your cursor back in the `print(x)` cell, and then press *Shift+Enter*.

   An error message appears because you haven't executed a statement that assigns a value to x in the restarted kernel.

3. The fourth cell is still empty. Type the following lines of code in that cell, and then press *Shift+Enter*:

```
x = 7
x = 99
x = 1024
print(x)
print(y)
```

With the execution of `print(x)`, the notebook displays 1024, but after that, you see an error message:

```
Traceback (most recent call last):
  Input In [2] in <cell line: 5>
    print(y)
NameError: name 'y' is not defined
```

The error message references *line 5*, and *line 5* is easy to find, but if the message tells you to find *line 29* among 54 lines, finding the line in question might not be so easy. To fix this problem, go to the menu bar at the top of the Jupyter notebook and select **View | Show Line Numbers**. When you do, line numbers appear in each cell of the notebook.

Tip

By using your mouse, you can drag cells from one position to another in a notebook. You can also add an empty cell after any existing cell. To do so, select an existing cell and click the plus sign icon in the notebook's toolbar.

Stopping a run of the code

When you use a Jupyter notebook, you're the boss. You start a run by pressing *Shift+Enter*, and you stop a run by interrupting the kernel. Try these steps:

1. In the empty cell, at the bottom of your notebook, type the following code, and then press *Shift+Enter*:

```
import time

while True:
    print('Hello')
    time.sleep(1)
```

This code throws Python into an infinite loop. Oops!

2. In the menu bar, at the top of the notebook, select **Kernel | Interrupt Kernel**. Doing so ends the infinite loop's run.

Saving your work

IBM's Jupyter notebook saves your most recent changes automatically, but it never hurts to issue an explicit **Save** command. Here's how you do it:

1. In the menu bar, near the top of the Jupyter window, select **File | Save Notebook As...**.

    A **Save File As...** pop-up box appears. The box contains a default filename such as `Untitled.ipynb`. The letters `ipynb` stand for the words **interactive Python notebook**.

2. In the pop-up box, replace whatever comes before `.ipynb` with some memorable name.

    Is `FirstNotebook.ipynb` memorable? If not, what about `JupyterDemo.ipynb`?

3. In the pop-up box, click on the **Save** button.

    Your notebook has a new, more informative name. You're ready to move on.

Copying this book's example code

In the previous sections, I encouraged you to start with an empty Jupyter notebook and type all the code yourself, but you don't always have to do this. You can download this book's Python code from the following GitHub link: `https://github.com/PacktPublishing/Quantum-Computing-Algorithms`. The code comes in two different formats, and you can decide which format suits your needs:

- **Markdown (.md) files**: Each markdown file contains code that you can copy and paste into your Jupyter notebook's cells. Each cell's code begins and ends with lines containing three backticks. For example, imagine a markdown file containing the following text:

```python
x = 1024
```

```python
x = 7
print(x)
```

    From this markdown file, copy `x = 1024` into one notebook cell, and copy the following lines into the next notebook cell:

```
x = 7
print(x)
```

- **Jupyter notebook (.ipynb) files**: Each such file encodes a complete Jupyter notebook – a notebook that's divided into one or more cells. To use one of these files, visit any Jupyter notebook page. Click the **View** menu item and make sure that the **Show Left Sidebar** sub-item is checked. In that left sidebar, click the upload icon (⬆). In the resulting dialog box, navigate to the .ipynb file you want to import. Then, click the dialog box's **Open** button.

In the next section, we use all our Jupyter notebook tricks to perform useful calculations.

Matrices in Python

Doing matrix operations by hand is tedious and error-prone, but you don't have to do everything by hand. This section describes a nice alternative:

1. If you haven't already done so, follow *steps 1 to 3* at the beginning of this chapter's *Jupyter notebooks* section.

2. In the upper-left corner of the page, click the **New File** button.

3. In the resulting **Launcher** tab, click the **Notebook | Qiskit** button.

    A new, empty Jupyter/Python notebook appears.

4. In the empty cell, near the top of the notebook, type the following code, and then press *Shift+Enter*:

    ```
    import numpy as np
    ```

    No output shows up on the screen, but behind the scenes, this statement makes a library named NumPy available to the code in your notebook. As the name suggests, NumPy deals with numbers and other mathematical concepts. From this point on, your code uses np as a nickname for NumPy.

Important note

The previous section (*Copying this book's code*) explains how you can download this book's code to avoid having to type it, but sometimes typing code yourself helps it sink into your brain. If you type the code yourself, remember to type every character exactly as you see it in this book. For example, in this cell's text field, the first word should be import (starting with a lowercase letter i), not Import.

---

5. In the next cell, type the following code, and then press *Shift+Enter*:

    ```
    A = np.matrix( [[2, -3, 0],
                    [1, 5, 19]] )
    print(np.dot(4, A))
    ```

    The notebook responds by displaying the following output:

    ```
    [[  8 -12   0]
     [  4  20  76]]
    ```

That's the answer we get when we do scalar multiplication in this chapter's *Working with matrices* section.

In NumPy, you enclose each row of a matrix in square brackets, and you enclose the rows themselves in an additional pair of square brackets (I've highlighted those additional brackets in this step's code and output). In the code, you don't have to put each row on a line of its own, but doing so makes your code more readable.

NumPy's `dot` function wears many hats. In this step, the `dot` function performs scalar multiplication.

6. In the next cell, type the following code, and then press *Shift+Enter*:

```
A = np.matrix( [[5, 1, 3]] )
B = np.matrix( [[2],
                [4],
                [6]] )
print(np.dot(A, B))
```

The notebook responds by displaying `[[32]]`. That's almost the answer we get when we compute a dot product in this chapter's *Vectors* section. With its strict adherence to its own row/column notation, NumPy replies by giving you a *1x1* matrix.

7. In the next cell, type the following code, and then press *Shift+Enter*:

```
A = np.matrix( [[1, 3, 2],
                [4, 0, -1]] )
B = np.matrix( [[5, 10],
                [2, -2],
                [-3, 6]] )
print(np.dot(A, B))
```

The notebook responds by displaying this:

```
[[ 5 16]
 [23 34]]
```

That's one of the answers we get in this chapter's *Matrix multiplication* section.

8. In the next cell, type the following code, and then press *Shift+Enter*:

```
print(A * B)
```

The notebook responds by displaying this:

```
[[ 5 16]
 [23 34]]
```

How nice! You can multiply matrices using an asterisk!

9. In the next cell, type the following code, and then press *Shift+Enter*:

```
A = np.matrix( [[3],
               [2]] )
B = np.matrix( [[5],
               [4]] )
print(np.kron(A, B))
```

The notebook responds by displaying this:

```
[[15]
 [12]
 [10]
 [ 8]]
```

That's one of the answers we get in this chapter's *Tensor product* section. The name **kron** stands for **Kronecker**, named after the 19th-century mathematician Leopold Kronecker.

10. Finish up by repeating the steps in this chapter's *Saving your work* section.

Summary

You can represent bits and bit sequences as vectors. You can represent operators on bits as matrices. To combine bits, combine operators, and apply operators to bits, you use matrix multiplication and the tensor product.

In truth, most people who deal with bit-level concepts don't do matrix calculations. There are other, less cumbersome ways to understand bits and bit operations, but quantum computing deals with things that are more complicated than bits. Quantum computing deals with quantum bits, also known as qubits, and matrices give us excellent insights into the workings of qubits.

The next chapter extends your understanding of matrices from bits to qubits. In that chapter, you begin investigating the weird world of quantum computing.

Questions

1. What's the output of the following circuit?

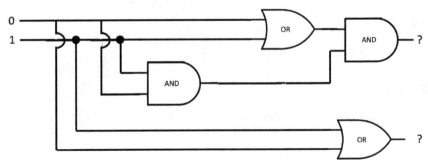

2. What's the decimal representation of the binary number 10111001?

3. Find the dot product:

$$\left(1 \ 0 \ 3 \ 2 \right) \begin{pmatrix} 5 \\ 8 \\ 4 \\ 4 \end{pmatrix}$$

4. Find the product of the two matrices:

$$\begin{pmatrix} 1 & 2 & 3 & 0 \\ 2 & 1 & -1 & 3 \end{pmatrix} \begin{pmatrix} 1 & 1 & -2 \\ 3 & 2 & -1 \\ 0 & 4 & 3 \\ 3 & -3 & 5 \end{pmatrix}$$

5. Find the tensor product:

$$\begin{pmatrix} 2 \\ 3 \\ 1 \end{pmatrix} \begin{pmatrix} 8 & 4 & 0 \\ 1 & 3 & 5 \end{pmatrix}$$

6. Express the following circuit in matrix form:

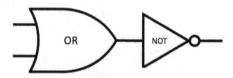

7. Express the following circuit in matrix form:

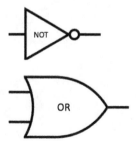

8. Redo *questions 4* and *5* using a Jupyter notebook with NumPy.

9. Look for the toolbar near the top of the Jupyter notebook. Hover over each of the toolbar's icons and read each tooltip that appears. These icons give you quick access to some of the notebook's more frequently-used commands.

2
What Is a Qubit?

"A **qubit** is a quantum bit." That's a nice soundbite, but what does it mean? In this chapter, we'll do our best to answer that question.

To describe a qubit, we'll start with a rough analogy. We'll rely on some very informal (maybe even heretical) terminology and work our way to a more precise description.

Our description will take several forms. We'll start by drawing a sharp distinction between qubits and classical bits. We'll assert that a qubit's value isn't necessarily 0 or 1. Instead, a qubit's value can be a number (of some kind) that's between 0 and 1. To make this assertion more believable, we'll explain how to implement qubits in a laboratory using elementary particles such as electrons and photons.

A bit can sit quietly on a thumb drive without ever being read by a computer of any kind. The bit's value is 0 or 1, whether anyone ever reads the bit or not. But to talk about a qubit's value, we have to consider a reading of that value. Discussing this concept will lead us to the role of measurement and probability in quantum mechanics.

All these ideas about qubits may seem vague and abstract. To make them concrete, we'll conclude this chapter by writing code using IBM's Qiskit quantum computing library.

We'll cover the following topics in this chapter:

- A qubit's values between 0 and 1
- Qubits and Qiskit
- Variations of this chapter's code

Technical requirements

Every chapter in this book requires that you have access to a web browser. You can download each chapter's Python code from the following GitHub link: `https://github.com/PacktPublishing/Quantum-Computing-Algorithms`.

A qubit's values between 0 and 1

A bit's value is either 0 or 1. In contrast, a qubit's value can be $|0\rangle$, $|1\rangle$, or anything in between $|0\rangle$ and $|1\rangle$. But that's not the whole story. Imagine a qubit that (in some way or other) is halfway between $|0\rangle$ and $|1\rangle$. This qubit has infinitely many different ways of being between $|0\rangle$ and $|1\rangle$. To understand this infinite number of ways, we need some visual aids.

Pictorially, each bit is on one of two points. *Figure 2.1* shows these points:

0
●

●
1

Figure 2.1 – A bit's value is either 0 or 1

In contrast, we can think of a qubit as being somewhere on the surface of a sphere. We call it a **Bloch sphere**. Here are some facts about the Bloch sphere:

- A qubit on the Bloch sphere's north pole has a value of $|0\rangle$. *Figure 2.2* shows this sphere:

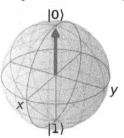

Figure 2.2 – A $|0\rangle$ qubit

- A qubit on the Bloch sphere's south pole has a value of $|1\rangle$. *Figure 2.3* illustrates this idea:

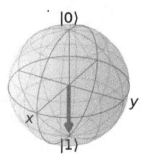

Figure 2.3 – A $|1\rangle$ qubit

- A qubit that's anywhere on the Bloch sphere's equator is (in some sense) halfway between $|0\rangle$ and $|1\rangle$:

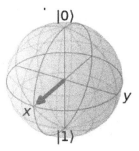

Figure 2.4 – A "halfway" qubit

We say that the qubit in *Figure 2.4* is in an equal **superposition** of the $|0\rangle$ and $|1\rangle$ states.

- Not all superpositions are equal superpositions. A qubit's value can be anywhere on the surface of the Bloch sphere. For example, the qubits in *Figure 2.5* are in three different superpositions of the $|0\rangle$ and $|1\rangle$ states:

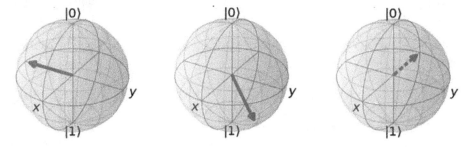

Figure 2.5 – Qubits on the sphere

In *Figure 2.5*, each sphere's arrow points from the center of the sphere to the sphere's outer surface. On the leftmost sphere, the arrow points above the equator, so this qubit isn't exactly halfway between $|0\rangle$ and $|1\rangle$. Instead, this leftmost qubit leans more toward being $|0\rangle$ than $|1\rangle$.

On the middle sphere in *Figure 2.5*, the arrow points below the equator. So, this qubit leans more toward being $|1\rangle$ than $|0\rangle$.

On the rightmost sphere in *Figure 2.5*, a dotted line indicates that the arrow points to a place on the far side of the sphere. That arrow points to a place along the equator, so the rightmost qubit is exactly halfway between $|0\rangle$ and $|1\rangle$.

But now, we have two "*halfway*" qubits – the qubit in *Figure 2.4* and the rightmost qubit in *Figure 2.5*. How do they differ? You'll start reading about this issue in *Chapter 3, Doing the Math*.

A qubit's value can be between $|0\rangle$ and $|1\rangle$. We've established this, but we've left many questions unanswered. Why is it useful for a qubit to be between $|0\rangle$ and $|1\rangle$? How do you make a quantum computer's qubit? Aside from the strange Bloch sphere business, what does being *between $|0\rangle$ and $|1\rangle$* mean? We will explore these issues in the next few sections.

> **Important note**
>
> In this chapter, I am using the phrase *between $|0\rangle$ and $|1\rangle$* because it paints a mental picture and isn't challenging to grasp. But, when you chat with a quantum computing professional, you should avoid using this phrase. What I call *between $|0\rangle$ and $|1\rangle$* in this chapter is a pair of numbers (possibly complex numbers) involving square roots and other mathematical tricks. For details, see *Chapter 3, Doing the Math*.

Are qubits useful?

Some authors refer to a qubit in superposition as *having both values, $|0\rangle$ and $|1\rangle$, at the same time*. Unfortunately, that simple picture can be misleading.

Instead of relying on the words *at the same time*, let's temporarily agree that a qubit in superposition *has characteristics of both $|0\rangle$ and $|1\rangle$*. It's vague wording, but it can be helpful.

Figure 2.6 contains a four-bit string, **0101**. If you interpret this string as a binary number, it represents the number five:

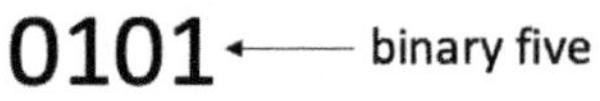

$$0101 \longleftarrow \text{binary five}$$

Figure 2.6 – One numeric value

Compare this with *Figure 2.7*. This figure contains four qubits, each of which is in a halfway state:

$${}^1 0 \quad {}^1 0 \quad {}^1 0 \quad {}^1 0 \qquad \longleftarrow \text{binary 0, 1, 2, ..., 15}$$

Figure 2.7 – 16 numeric values

A string of four qubits has characteristics of 0101, 0000, 1111, and all the other combinations of four zeros or ones. That's 16 combinations in all – a lot of information for only four measly qubits!

And it gets even better. If you add another qubit, you have five qubits, representing 32 combinations of zeros and ones. Every time you add a qubit, you double the number of combinations. This is exponential growth. On your laptop computer, a string of 64 bits represents only one number. But, on a quantum computer, a string of 64 qubits has characteristics of 18,446,744,073,709,551,616 different numbers!

Of course, the devil is in the details. Sifting a correct answer out of all these combinations is no simple task. It's the subject of years of research on the part of quantum computing professionals.

How to make a qubit

The previous section treated qubits as abstract entities. That's nice, but you can't do much with qubits if you don't know to build them. So, engineers around the world are investigating ways to make qubits. Here are some of the things they've tried:

- A **photonic qubit** is a single particle of light.

  Companies that support photonic qubit development include Xanadu and PsiQuantum.

- A **trapped ion qubit** is the outermost electron of an ion, such as Calcium+ or Strontium+.

  Honeywell and IonQ have quantum computers with trapped ion qubits.

- A **superconducting qubit** is a resistance-free current at a very low temperature.

  Some big players develop hardware with superconducting qubits. This includes Intel, Google, D-Wave, and Rigetti. The examples you will see in this book come courtesy of IBM, whose quantum computers have superconducting qubits.

- A **topological qubit** is a collection of quasiparticles – phenomena that behave as elementary particles even though they're not really elementary particles.

  You may have heard of a company named Microsoft. That company works with topological qubits.

At a simplified level, photons and trapped ions are fairly easy to visualize. They help us understand how qubits work; we'll explore them in the next two sections.

Photons

A single particle of light is called a **photon**. A photon can have various kinds of **polarization**, such as **linear polarization** and **circular polarization**, both of which are specific kinds of **elliptical polarization**. In this section, we'll only consider linear polarization.

When you catch a linearly polarized photon, that polarization is either **horizontal** or **vertical**. *Figure 2.8* illustrates this idea:

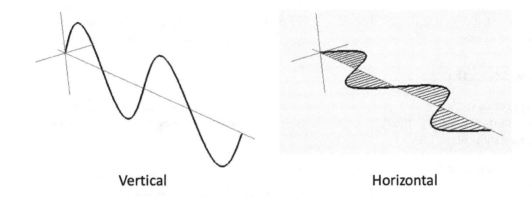

Vertical   Horizontal

Figure 2.8 – Linear polarization

To implement qubits using photons, we must agree on what it means to be $|0\rangle$ versus $|1\rangle$. For example, we can consider vertical polarization to be $|0\rangle$ and horizontal polarization to be $|1\rangle$. It doesn't matter which is which, so long as we're consistent.

Take a moment to think about London's Big Ben clock tower. Is it vertical or horizontal? If you're standing in Parliament Square and staring at the tower, the tower is vertical. But if you're lying on a park bench and looking in the same direction, the tower is horizontal. In that sense, photons behave the same way. A photon alone is neither vertical nor horizontal. A photon can be *horizontal concerning an agreed-upon viewing angle* or *vertical concerning an agreed-upon viewing angle*. When we describe photons, we often omit the tedious *agreed-upon angle* phrase. But, no matter what we omit, a photon's polarization is in the eye of the beholder. To better understand this idea, consider the following examples:

- *Figure 2.9* shows a filter whose openings are vertical lines. When a vertically polarized photon hits the vertical filter, the photon goes right through:

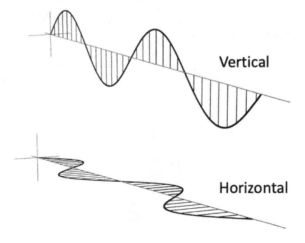

Vertical

Horizontal

Figure 2.9 – A vertically polarized photon meets a vertically oriented filter

- *Figure 2.10* shows a filter whose openings are horizontal lines. When a horizontally polarized photon hits the horizontal filter, the photon goes right through:

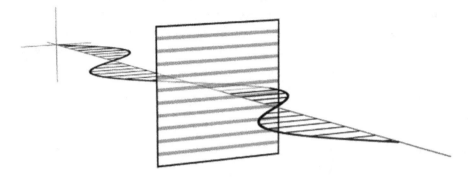

Figure 2.10 – A horizontally polarized photon meets a horizontally oriented filter

- What if a vertically polarized photon hits a horizontal filter? In that scenario, the filter's horizontal barrier blocks the photon. The photon doesn't get through. *Figure 2.11* shows this:

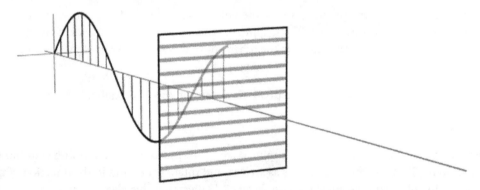

Figure 2.11 – A vertically polarized photon meets a horizontally oriented filter

What about diagonal orientations? Near the start of this section, you traveled to London to see the Big Ben clock tower. Now, you fly from England to Italy and visit the Leaning Tower of Pisa. When you look at the tower, you see it leaning diagonally.

But the tower isn't a photon. If it were, you would see something completely different.

Imagine a visit to the Leaning *Photon* of Pisa. When you look toward the Photon, you may or may not see it. If you visit the Photon several days in a row, you will see it on some days and not on others. On the days when you see it, the Photon *is aligned along whatever way you're standing*. Unlike the famous Tower of Pisa, a photon never appears to be leaning at an angle.

Here are some examples:

- Imagine a filter whose openings are oriented diagonally. What if a vertically polarized photon hits this diagonal filter? The surprising answer is, the photon has a 50/50 chance of getting through. A photon that encounters a diagonally oriented filter is in a half-$|0\rangle$-half-$|1\rangle$ state. What's more, if the photon gets through, it becomes polarized along the filter's diagonal orientation. *Figure 2.12* illustrates this concept:

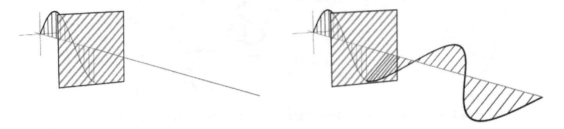

Figure 2.12 – A vertically polarized photon meets a diagonally oriented filter

- Imagine a filter whose openings are oriented diagonally. What if 1,000 vertically polarized photons hit this diagonal filter? Then, approximately 500 of them are blocked by the filter. The others pass through the filter. On the other side of the filter, these photons have the same orientation as the filter.

So much for photonic qubits. Let's move on to the electrons in trapped ion qubits.

Electrons

We can implement a qubit with an electron's spin. Make a fist with your right hand with your thumb pointing upward. Think of your fingernails as arrows, and rotate your hand in the direction of the arrows. That kind of rotation is called **spin up**. *Figure 2.13* illustrates this idea:

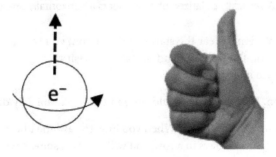

Figure 2.13 – Spin up

Once again, rotate your right hand in the direction of your fingertips. But this time, start with your thumb pointing down. That kind of rotation is called **spin down**. *Figure 2.14* brings this point home:

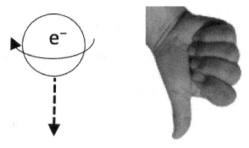

Figure 2.14 – Spin down

When you apply a magnetic field to an electron, the electron moves in one of two directions, depending on the electron's spin. That's how you measure spin up versus spin down.

Now, imagine that you have 1,000 spin-up electrons and you are measuring their spins with an instrument that's tilted on its side. *Figure 2.15* shows you a few of these electrons:

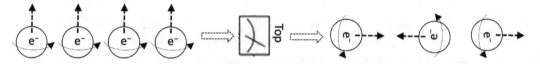

Figure 2.15 – A spin-up electron meets a measuring device that's tilted sideways

After the measurement, roughly half of them will have a spin in the same sideways direction as the measuring instrument. The rest of them will have spins in the opposite sideways direction as the measurement.

Tilt your head sideways so that, in *Figure 2.15*, the word *Top* appears to be on top of the measuring device, and look at the three electrons that are moving away from the device. From this sideways point of view, two of those three electrons spin up, and the remaining electron spins down.

To implement qubits using electrons, we must agree on what it means to be $|0\rangle$ versus $|1\rangle$. For example, we can consider spin up to be $|0\rangle$ and spin down to be $|1\rangle$. Alternatively, we can let spin up represent $|1\rangle$ and spin down represent $|0\rangle$. It doesn't matter which choice we make, so long as we're consistent. But notice how electron spin differs from photon polarization. With photon polarization, $|0\rangle$ and $|1\rangle$ differ by 90 degrees. With electron spin, $|0\rangle$ and $|1\rangle$ differ by 180 degrees.

Before leaving the subject of qubit implementations, we must make one thing clear: Photons don't wave in a direction, and electrons don't spin around. Words such as *wave* and *spin* help us remember some of the properties of these physical phenomena, but they don't convey the whole picture. In fact, with our everyday experiences being on a macroscopic scale, we can't convey the whole picture.

The only precise way to describe quantum mechanics is with mathematics. In the next section, we'll take our first peek at the mathematics of probability.

What does "between |0⟩ and |1⟩" mean?

If someone says *halfway between black and white*, you imagine an even mix of the two shades, also known as *gray*. If someone says *halfway between comedy and drama*, you may imagine a patchwork in which some scenes are funny, and others are serious. So, what do I mean when I loosely describe a qubit as being *between |0⟩ and |1⟩*? Is it an even mix such as black and white or a mosaic of scenes such as comedy and drama? Unfortunately, it's neither of these. Qubits don't behave like anything we experience in daily life.

> **Important note**
>
> Quantum physics is all about observations and measurements. You **observe** a qubit by **measuring** the qubit's state. But here's the catch: when you measure a qubit, you irreversibly force the qubit to be either 0 or 1. This happens even when the qubit is in a halfway state before the measurement. To say that a qubit is in a halfway state is to assert that, when you measure the qubit, the probability of observing 0 is equal to the probability of observing 1.

I urge you to read that paragraph a few times. According to those sentences, the halfway state of a qubit is defined in terms of *measurement* and *probability*. If a qubit falls in the proverbial forest, and there's no one around to measure it, the qubit has no definite state. That's one of the bizarre things about modern physics.

Since the words *measurement* and *probability* are so important, we better make sure that we know what they mean.

About measurement

To *measure* something is to interact with that thing. When a measuring device interacts with something, the device *discovers* something as a result of the interaction.

When we have two measuring devices, we should find out whether they're calibrated the same way. For example, if you say that your room is 10 meters wide, and the carpet installer thinks you said *10 yards wide*, you're going to get the wrong size carpet. Your meter stick isn't calibrated the way the installer's yardstick is calibrated. It's the same with quantum measurements. If my filter for photons points upward from the ground, and your filter points sideways, we're going to get different results.

So, let's make a deal: in a particular experiment, unless we say otherwise, all qubit measurements are done from the same point of view. If we're measuring polarized light, everyone's filter points in the same direction. If we're measuring electron spin, everyone agrees on what *spin up* and *spin down* mean.

A fascinating side note

One way to think about quantum physics is to imagine that anything can measure anything else. A measurement takes place when one thing interacts with another. In a sense, the measurement is more real than the things that interact. When a tree falls in a forest and no one hears it fall, the tree and the ground measure one another. If you think about this too much, you'll go crazy!

About probability

What does the word *probability* mean? That's a tough one. Scholars throughout history have named at least four interpretations of the notion of probability:

- **Frequentist**: If X happens more often than Y, then X is more probable than Y

- **Classical**: If the number of ways X can happen is more than the number of ways Y can happen, then X is more probable than Y

- **Subjective**: If X is more believable than Y, then X is more probable than Y

- **Propensity**: If X has stronger causes than Y, then X is more probable than Y

In quantum computing, we use the frequentist interpretation.

Important note

Here's what it means for a qubit to be *halfway between $|0\rangle$ and $|1\rangle$*: if you measure many qubits, each of which is in the same *halfway* state, approximately half of those qubits will be 0, while half of them will be 1.

In the previous paragraph, my use of the word *many* is intentional. To describe what it means for a single qubit to be in a halfway state, you have to talk about measuring many qubits that are all in the same halfway state. This is a strange, counterintuitive way of describing *halfway*. It means that you can never decide that a single qubit is *halfway*. Instead, you have to measure many identical qubits and find the ratio of 0 measurements to 1 measurements. There's nothing like this in our common, everyday experience.

If qubits were alive, they'd be more like ants than humans. They'd be the ultimate practitioners of collective behavior. The only way we can learn about a qubit is to measure many qubits that are exactly like it. Measuring a single qubit tells us almost nothing about that particular qubit.

To help bring this point home, let's do a few thought experiments:

- You send a qubit to Bob. Bob receives the qubit and says, "Based on my measurement, I conclude that this qubit was in the $|1\rangle$ state." How should you respond?

  Tell Bob that he's full of baloney. It's impossible to determine a single qubit's state by measuring the qubit. The only qubit that could never be 1 after a measurement is a qubit whose state is

100% $|0\rangle$. Any other qubit – a halfway qubit or a *quarter-way* qubit (whatever that is) – can look like 1 when you measure it.

- You prepare a qubit in the halfway state and send that qubit to Bob. Bob receives the qubit and says, "Based on my measurement, I conclude that this qubit was in the halfway state." How should you respond?

  Tell Bob to take a hike. When Bob measures a single qubit, he sees 0 or 1. He never sees anything in between $|0\rangle$ and $|1\rangle$.

- Alice's and Bob's measuring devices are oriented the same way. You prepare a qubit in the halfway state and send that qubit to Bob. Bob receives the qubit and says, "I measured your qubit and got 1." Immediately afterward, Bob gives the qubit to Alice, who says, "I measured the qubit and got 0." How should you respond?

  Tell Bob and Alice that one of them is lying. Assume, for the moment, that Bob correctly got 1 when he measured the qubit. If so, that qubit stays in the $|1\rangle$ state until someone (or something) performs an operation on the qubit. If no operation occurs, Alice receives a qubit that's 100% in the $|1\rangle$ state. So, Alice observes the value 1.

- You prepare a qubit in the halfway state and send that qubit to Bob. Bob receives the qubit and says, "I measured your qubit and got 1." Bob puts that same qubit back into a halfway state and then gives the qubit to Alice. Alice says, "I measured the qubit and got 0." How should you respond?

  Thank Bob and Alice for their accurate reporting. By measuring the qubit and observing the value 1, Bob put that qubit into the $|1\rangle$ state. But then Bob performed an operation to put that qubit back into a halfway state. So, Alice may still measure either a 0 or a 1.

- You prepare 100 qubits, each in the halfway state, and send those qubits to Bob. Bob receives the qubits and says, "When I measured the qubits, 48 of them were 0, while 52 were 1." How should you respond?

  Send Bob some money for completing his task correctly. Quantum phenomena are statistical. If you measure 100 qubits, each in the same halfway state, you'll get *approximately* 50 0 measurements and 50 1 measurements. There's no guarantee that you'll get exactly 50 of each.

In summary, qubits don't behave like common, everyday objects. Measuring a single qubit's value tells you almost nothing about the qubit. By measuring many qubits, you discover their collective, probabilistic properties.

With all these ideas about qubits under your belt, you can begin writing some quantum computing code.

Qubits and Qiskit

Chapter 1, New Ways to Think about Bits, introduced Jupyter notebooks and showed you how to multiply matrices in Python. In this chapter, we'll up the ante with Qiskit code for qubits. We'll write code to create qubits, modify qubits' values using quantum operators, measure qubits, and then display the results.

Creating and running a quantum circuit

To create your first quantum computing program, follow these steps:

1. Create a new Jupyter notebook by following *Steps 1* through *3* in the *Matrices in Python* section in *Chapter 1, New Ways to Think about Bits.*

2. In the cell at the top of the notebook, copy the following code, and then press *Shift + Enter*:

```
from qiskit import QuantumRegister, \
    ClassicalRegister, QuantumCircuit

qReg = QuantumRegister(1, 'q')
cReg = ClassicalRegister(1, 'c')
circuit = QuantumCircuit(qReg, cReg)
circuit.h(qReg[0])
circuit.measure(qReg[0], cReg[0])
display(circuit.draw('latex'))
```

Our GitHub link (`https://github.com/PacktPublishing/Quantum-Computing-Algorithms`) contains code that you can import as a Jupyter notebook or copy and paste into a notebook's cells. Alternatively, you can type the code yourself.

If you type the code yourself, remember to type every character exactly as you see it in this book. The first word should be `from` (starting with a lowercase letter `f`), not `From`. Type `QuantumCircuit`, not `quantumcircuit` or `Quantum Circuit`. Don't indent any of the lines. Type `'latex'` (using straight quotes), not `'latex'` (with so-called *smart quotes*).

If you don't see a response, keep your eye on the page's status bar. Among other things, you'll see the word `Busy` or the word `Idle`. The word *Busy* means you should wait longer for Qiskit's response. The word *Idle* means Qiskit has finished doing whatever it intends to do.

When Qiskit becomes idle, it responds by displaying the diagram shown in *Figure 2.16*. This diagram shows the circuitry that you've defined by calling constructors and methods in your code:

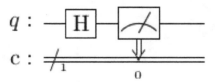

Figure 2.16 – The resulting circuit

The circuit in *Figure 2.16* consists of one qubit (labeled *q*) and one classical bit (labeled *c*). The qubit flows from left to right, encountering two gates (*H* and then a measuring device) along the way. For a more thorough explanation of the code in this step, see this chapter's *Creating a circuit* section.

After the diagram in *Figure 2.16* appears on your screen, the notebook will create a new cell with an empty text field.

3. In the new cell's text field, copy the following code, and then press *Shift + Enter*:

```
from qiskit import Aer
device = Aer.get_backend('qasm_simulator')
```

Roughly speaking, a **backend** is a computer that can run the circuit that you've created. For details, see this chapter's *Finding a backend* section.

4. In a new empty cell, copy the following code, and then press *Shift + Enter*:

```
from qiskit import execute
from qiskit.visualization import plot_histogram

job = execute(circuit, backend=device, shots=1000)
print(job.job_id())

result = job.result()
counts = result.get_counts(circuit)

print(counts)
display(plot_histogram(counts))
```

The purpose of this code is to implement your circuit on a quantum computer or simulator and feed the circuit 1,000 qubits. The code counts the number of times the circuit's measuring device registers 0 versus 1 and displays the count in two different ways. For details, see this chapter's *Running the circuit* section.

When the page's status bar says *Idle*, you'll see the output shown in *Figure 2.17* (or something very much like it):

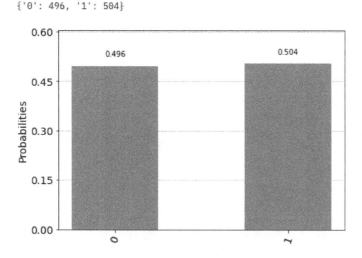

`{'0': 496, '1': 504}`

Figure 2.17 – The output of the code in Step 4

Qiskit sent 1,000 qubits through your circuit and measured the output each time. The small print stating `{'0': 496, '1': 504}` at the very top of *Figure 2.17* says that, out of the 1,000 qubits sent through this circuit, the computer measured 0 exactly 496 times and 1 exactly 504 times. The chart shows this information in graphical form.

For more detailed explanations of the code and its output, see this chapter's *Running the circuit* section.

5. With your cursor in the code from *Step 4*, press *Shift + Enter* a second time.

    Once again, you will see a measurement count and a chart. But this time, you will see different numbers. Maybe, instead of `{'0': 496, '1': 504}`, you see `{'0': 530, '1': 470}`. Quantum mechanics is statistical. It's a fact of nature.

The next section provides more details about the code in your Qiskit notebook.

Understanding the Qiskit code

In the *Creating and running a quantum circuit* section, I asked you to copy some code, run it, and then examine its output. If all went well, you should have ended up with that warm, fuzzy feeling of having successfully run a program. But did you understand the program? Maybe not. In this section, we'll explore some of the key concepts regarding the design of your quantum computing code.

Creating a circuit

There are a few ways to design quantum computers. The most common among them is **gate-based quantum computing** (also known as **universal quantum computing**). With this model, you define one or more wires, each with its own qubit. A wire's qubit interacts with one or more gates. Each gate performs an operation of some kind. At the end of the wire, you measure the qubit, always getting either 0 or 1. You deposit the measurement result in yet another wire (a wire that's reserved for a classical bit).

> **Disclaimer**
>
> In this book, what I call a "wire" isn't necessarily a thin piece of metal for transporting electrical signals. Instead, a **wire** is a metaphorical name for the transitions that a qubit (or bit) undergoes from one operation to the next. Many authors avoid the word "wire" in favor of the terms "qubit" or "classical bit." I've tried to get out of the habit of writing "wire" but, alas, I can't do it. I can't stop writing phrases like "a qubit goes through a gate" even though gates aren't physical things that qubits move through. Sorry about that.

Here's a copy of the code in *Step 2* of the *Creating and running a quantum circuit* section:

```
from qiskit import QuantumRegister, \
    ClassicalRegister, QuantumCircuit    # 1

qReg = QuantumRegister(1, 'q')           # 2
cReg = ClassicalRegister(1, 'c')         # 3
circuit = QuantumCircuit(qReg, cReg)     # 4
circuit.h(qReg[0])                       # 5
circuit.measure(qReg[0], cReg[0])        # 6
display(circuit.draw('latex'))           # 7
```

I've numbered the code's statements for easy reference. Refer back to the code's statement numbers as you read the explanations that follow:

- Statement 1 makes Qiskit's `QuantumRegister`, `ClassicalRegister`, and `QuantumCircuit` modules available for use in this notebook. A `QuantumRegister` is a collection of one or more qubits, while a `ClassicalRegister` is a collection of one or more bits.

- Statement 2 creates a `QuantumRegister` consisting of only one wire. It labels the register with the letter q and assigns the register to the qReg variable. *Figure 2.18* shows how Qiskit might draw this register:

$$q: \quad \underline{\qquad}$$

Figure 2.18 – One qubit in the |0⟩ state

The one and only wire on this register is called `qReg[0]`. If this register had three wires, they'd be called `qReg[0]`, `qReg[1]`, and `qReg[2]`. Each wire hosts a single qubit, and each qubit starts in state |0⟩.

- Statement 3 creates a `ClassicalRegister` consisting of only one wire. It labels the register with the letter c and assigns the register to the `cReg` variable. *Figure 2.19* shows how Qiskit might draw this register:

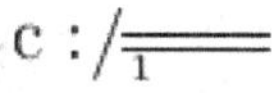

$$c : /\!\!=\!\!=_{\overline{1}}$$

Figure 2.19 – A classical register

A classical register hosts bits, not qubits. So, the values along a classical register can be 0s and 1s – no halfway states.

The one and only wire on this register is called `cReg[0]`.

When Qiskit draws a classical register, the default is to bundle the register's wires into one double line. In *Figure 2.19*, the slash with a number indicates the count of wires on the classical register. For details, see the *Drawing circuits* section.

- Statement 4 combines `qReg` and `cReg` to form a `QuantumCircuit`. *Figure 2.20* shows the circuit with both registers:

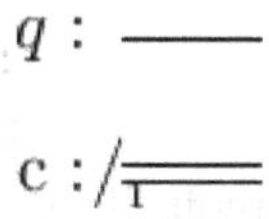

$$q : \quad \underline{\hspace{1cm}}$$

$$c : /\!\!=\!\!=_{\overline{1}}$$

Figure 2.20 – Combining two registers

Why do you need to combine registers into a circuit? Quantum computing involves two kinds of operations – **single-gate operations** and **multi-gate operations**. A single-gate operation modifies the state of a single qubit on a single wire. That's no big deal. But a multi-gate operation modifies one wire's qubits based on the properties of other wires' qubits. For the wires to communicate with one another, they must be part of the same circuit.

Notice that a `QuantumCircuit` can have both classical registers and quantum registers. Any quantum circuit worth defining has both kinds of registers. The quantum registers do calculations with qubits, and the classical registers store the results after measurements.

- Statement 5 puts an h gate on the wire whose index is 0 in `qReg` (the only wire in `qReg`). *Figure 2.21* shows the circuit with this h gate added:

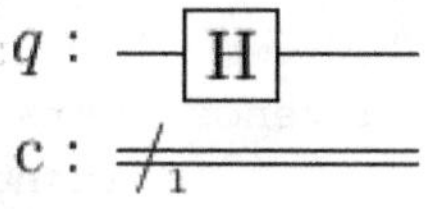

$$q : \quad \boxed{\text{H}} \underline{\hspace{1cm}}$$

$$c : \quad /\!\!=\!\!=_{\overline{1}}$$

Figure 2.21 – Adding a Hadamard gate

In statement 5, the letter h stands for **Hadamard**, named after mathematician Jacques Hadamard. When a |0⟩ qubit passes through a Hadamard gate, that qubit goes into the half-|0⟩-half-|1⟩ state. As you'll see, a Hadamard gate is the workhorse of quantum computing.

- Statement 6 measures the one and only qubit on the qReg register (the qubit whose index is 0). The statement places the measurement's result on the cReg wire whose index is 0. *Figure 2.22* shows the circuit that you've created:

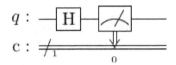

Figure 2.22 – Measuring a qubit

In *Figure 2.22*, the little **0** beneath the classical register stands for 0 in qReg[0].

> **Important note**
> A classical register can store only 0s and 1s. That's okay because measuring a qubit always yields either 0 or 1.

- Statement 7 displays a drawing of the circuit (the drawing in *Figure 2.22*).

  For the diagrams in this book, I have used the 'latex' drawing style, but many other styles are available (visit https://qiskit.org/documentation/stubs/qiskit.circuit.QuantumCircuit.draw.html for details).

Finding a backend

As mentioned previously, a **backend** is a computer that { XE "backend"} can run the circuit that you've created. When you run the code in Step 3 of the Creating and running a quantum circuit section, you make the device variable refer to a certain backend. This particular backend (named qasm_simulator) belongs to a family of backends known as **Aer**.

A **provider** is an agent that can find backends for you to use. When you run the following code, a list of backends will appear.

```
from qiskit_ibm_provider import IBMProvider

provider = IBMProvider()
IBM_cloud_backends = provider.backends(operational=True,
                                        min_num_qubits=5)
for i in IBM_cloud_backends:
    print(i)
```

With IBMProvider, some backends have the word simulator in their names. The one with `simulator` in their names aren't quantum computers. They're classical computers that do calculations to mimic a quantum computer's behavior. These classical computers do the same calculations you would do using matrices and matrix operations. They can mimic quantum computers only because, in your circuit, the number of qubits is so small.

Any `IBMProvider` backend that doesn't have `simulator` in its name is a real quantum computer. If you run your circuit on a real quantum computer (instead of a simulator), the run will take longer. (Of course, running on a real quantum computer exposes important issues that don't arise with runs on a simulator. For some details, see the discussion about decoherence in *Chapter 3*.) To compare pricing for simulator and quantum computer runs, check the IBM Quantum website (`https://www.ibm.com/quantum`). If you want, you can even run Qiskit programs with providers other than IBM. For example, `https://docs.rigetti.com/qcs/references/rigetti-provider-for-qiskit` contains information on running programs with Rigetti hardware.

> **Important note**
>
> In this chapter, we assume that your Jupyter notebook lives on the IBM cloud. If you want, you can run the notebook and a copy of Qiskit on your computer. For setup instructions, see `https://qiskit.org/documentation/getting_started.html`.

Running the circuit

In the *Creating and running a quantum circuit* section, the code in *Step 2* created a small circuit with an h gate. The h gate puts the circuit's one and only qubit into a *halfway between |0⟩ and |1⟩* state.

Your next task is to send qubits through the circuit. The code in *Step 4* of the *Creating and running a quantum circuit* section does just that.

Here's a copy of the code in *Step 4* of the *Creating and running a quantum circuit* section:

```
from qiskit import execute
from qiskit.visualization import plot_histogram

job = execute(circuit, backend=device, shots=1000)     # 1
print(job.job_id())

result = job.result()                                  # 2
counts = result.get_counts(circuit)                    # 3

print(counts)                                          # 4
display(plot_histogram(counts))                        # 5
```

Refer back to the code's statement numbers as you read the explanations that follow:

- In statement 1, the `execute` call runs the circuit on a backend and returns information about the run.

  That information is assigned to the `job` variable. The information includes an identification number, the job's creation date and time, and other useful data.

  In *Figure 2.22*, a qubit first passes through an h gate. So, as it approaches the measurement gate, this qubit is in a half-|0⟩-half-|1⟩ state. But when you measure a qubit, you get either 0 or 1 – never anything in between. So, the measuring of a single qubit doesn't provide much information. To get any real insight, you have to run the circuit many times and determine the ratio of 0 measurements to 1 measurements. Each of these runs is called a **shot**. The `shots=1000` parameter in statement 1 tells the computer to perform 1,000 runs.

- Statement 2 filters particular information from the aforementioned `job` variable. In particular, the job's `result` includes information such as the amount of memory used, the amount of time taken, and tallies of the measurements from all the shots.

- In statement 3, the `get_counts` method sifts the measurement tallies from the big `result` object.

- Statements 4 and 5 create the output shown in *Figure 2.17*. The code's `print` call shows the number of 0s and the number of 1s measured. The `plot_histogram` call shows the same information in a graphical format. The number at the top of each bar is a **proportion** – a number from 0.000 to 1.000.

What I mean by the word *proportion* is described best with a few examples:

- An event with a proportion of 0.000 never takes place.

- An event with a proportion of 1.000 always takes place.

- An event with a proportion of 0.496 takes place 496 out of 1,000 times. If you change the number of shots to 10,000, the event takes place 4,960 out of 10,000 times, and so on.

Proportions don't have to have exactly three digits to the right of the decimal point, but in this example, Qiskit gives us three digits.

Proportions are almost the same as probabilities. I often write *probabilities* when I mean *proportions*. For details about all this, see *Chapter 3*.

Variations of this chapter's code

This book doesn't cover all the features of Qiskit's extensive library, but it does cover enough of Qiskit to help you implement some important quantum computing algorithms. To this coverage, we have added some discussion of features that make Qiskit coding easier. Qiskit provides a dozen ways to accomplish any particular task, and this book describes two or three of them.

In the previous section, you learned one way to create and display quantum circuits. In this section, you'll learn a few more tricks.

Drawing circuits

Consider the following code:

```
from qiskit import QuantumRegister, \
    ClassicalRegister, QuantumCircuit

qReg = QuantumRegister(2, 'q')
qRegNew = QuantumRegister(2, 'qNew')
cReg = ClassicalRegister(2, 'c')
cRegNew = ClassicalRegister(1, 'cNew')
circuit = QuantumCircuit(qReg, qRegNew, cReg, cRegNew)
display(circuit.draw('latex'))
```

This code defines two quantum registers and two classical registers:

- The qReg variable refers to a 2-qubit register that will be labeled with the letter q.
- The qRegNew variable refers to a 2-qubit register that will be labeled with the name qNew.
- The cReg variable refers to a 2-bit register that will be labeled with the letter c.
- The cRegNew variable refers to a 1-bit register that will be labeled with the name cNew.

Figure 2.23 shows what you'll get when you run this code:

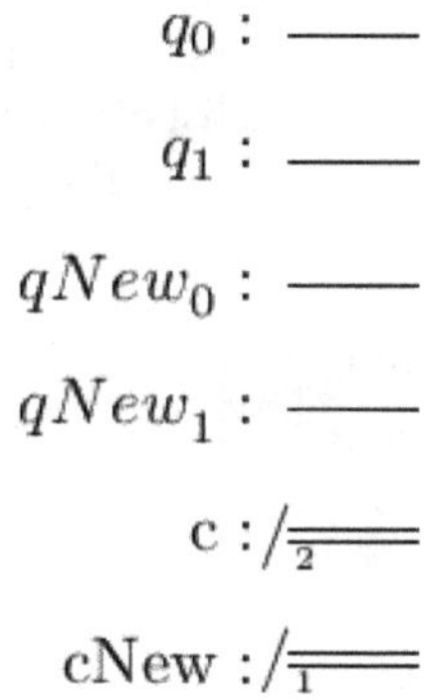

$$q_0 : \text{------}$$

$$q_1 : \text{------}$$

$$qNew_0 : \text{------}$$

$$qNew_1 : \text{------}$$

$$c :/\!\!\!=_{\overline{2}}$$

$$cNew :/\!\!\!=_{\overline{1}}$$

Figure 2.23 – Two quantum registers and two classical registers

In *Figure 2.23*, each quantum qubit is labeled with the name of its register and a numerical index.

Notice how Qiskit draws the classical registers. The default is to bundle each register's wires into one double line. Each classical bit is grouped with any other bits in the same classical register. For each classical register, a slash with a number indicates the count of wires on that register.

If you don't like the way *Figure 2.23* looks, you can change the `draw` method's settings. For details, visit https://qiskit.org/documentation/stubs/qiskit.circuit.QuantumCircuit.draw.html.

Creating a circuit with very little code

Up to this point in this chapter, the code to create a circuit has included every detail regarding the circuit's construction. In many programs, this much detail is unnecessary. For example, the code that follows is a much less verbose way to create the circuit in *Step 2* of the *Creating and running a quantum circuit* section:

```
from qiskit import QuantumCircuit

circuit = QuantumCircuit(1, 1)
circuit.h(0)
circuit.measure([0], [0])
display(circuit.draw('latex'))
```

The circuit that's created by this code is identical to the circuit in *Figure 2.16*. Here's a brief explanation of the code:

- The `QuantumCircuit(1, 1)` constructor call creates one quantum wire and one classical wire
- The call to `circuit.h(0)` adds an h gate to the quantum wire whose index is 0
- The call to `measure([0], [0])` adds a measurement from the 0-index quantum wire to the 0-index classical wire

In Qiskit's `measure` method, the first parameter is a list of quantum qubits, and the second parameter is a list of classical bits. (I tend to forget this fact when the parameters are lists of numbers). The following example shows how this works:

```
from qiskit import QuantumCircuit

circuit = QuantumCircuit(2, 4)
circuit.h(0)
circuit.barrier()
circuit.measure([1, 0], [2, 3])
display(circuit.draw('latex'))
```

Figure 2.24 shows the circuit generated from this code:

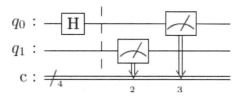

Figure 2.24 – A circuit with two quantum wires and four classical wires

In this code, the `circuit.measure([1, 0], [2, 3])` call tells Qiskit to do the following:

- Measure qubit 1 and put the result in classical wire 2

- Measure qubit 0 and put the result in classical wire 3

In this code, a call to `barrier` tells Qiskit not to mix the measurements with any gates that come before them. Without this `barrier`, you'd get the circuit shown in *Figure 2.25*:

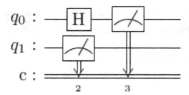

Figure 2.25 – A Hadamard and measurement may occur simultaneously

If you want to avoid the `measure` method's finicky behavior, you can call `measure_all` instead. Here's an example:

```
from qiskit import QuantumCircuit

circuit = QuantumCircuit(2)
circuit.h([0, 1])
circuit.measure_all()
display(circuit.draw('latex'))
```

Here's what's going on in this code:

- The `QuantumCircuit(2)` constructor call creates two quantum wires and no classical wires.

- The `circuit.h([0, 1])` call puts Hadamard gates on the 0-index wire and the 1-index wire.

- The call to `measure_all` creates two classical wires and two measurement gates. By default, the `measure_all` call puts a barrier before the measurement gates.

The result is the circuit shown in *Figure 2.26*:

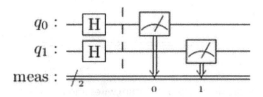

Figure 2.26 – Four wires and four gates from very few lines of code

When you do gate-based quantum computing, you create circuits. A circuit consists of registers, each consisting of either qubits or classical bits. Each qubit passes through a sequence of gates, the last of which is typically a measurement gate. The result of each measurement is a classical bit.

Summary

A qubit is like a bit, except that a qubit's value can be *between* |0⟩ and |1⟩. What *between* means requires more explanation but, in this chapter, we described *between* in a statistical fashion.

There's no way for us to determine that a single qubit's state is between |0⟩ and |1⟩. Instead, being *halfway between |0⟩ and |1⟩* means that, if you measure many such qubits, you get 0 approximately half the time and 1 all the other times. All we can do is record these measurement tallies. As my friend and colleague, Ashley Carter, once said, "*Quantum mechanics isn't about what "is." Quantum mechanics is about what we can measure.*"

In the next chapter, we'll use mathematics to describe qubits more precisely.

Questions

Alice stands up straight while she performs experiments with photons and filters. In each of the experiments, *vertical* refers to the way Alice is standing, and *horizontal* refers to the line of the floor:

1. In the first experiment, 1,000 vertically polarized photons hit a vertically polarized filter. How many of the photons get through the filter?

2. In the next experiment, 1,000 vertically polarized photons hit a horizontally polarized filter. How many of the photons get through the filter?

3. In Alice's next experiment, 1,000 vertically polarized photons hit a diagonally polarized filter. The ones that make it through hit another filter that's diagonally polarized in the same direction as the first filter. Assuming that 500 photons made it through the first filter, how many of them make it through the second filter?

4.   In another experiment, 1,000 vertically polarized photons hit a diagonally polarized filter. The ones that make it through hit another filter that's polarized horizontally. Assuming that 500 photons made it through the diagonal filter, how many of them make it through the horizontal filter?

5.   Run the code in *Step 4* of this chapter's *Creating and running a quantum circuit* section several times. Notice how the number of 0s and 1s in the output changes each time. Is the output ever exactly 500 zeros and 500 ones? How far does the output vary from that 500:500 match?

6.   Modify the code in the *Creating a circuit with very little code* section so that you get the circuit diagram shown in *Figure 2.27*:

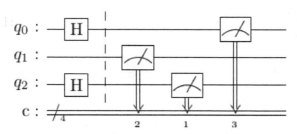

Figure 2.27 – Three qubits and three classical bits

3
Math for Qubits and Quantum Gates

In the previous chapter, we examined the idea of a qubit with its elusive behavior and its probabilistic nature. That chapter mentioned states other than $|0\rangle$, $|1\rangle$, and a *halfway* state but doesn't describe any details about such states.

To understand the full range of states a qubit can have, we need some math, and this chapter lays the mathematical groundwork. We'll start this chapter by replacing the intuitive *halfway* terminology with numbers. We'll show how those numbers apply to a qubit's state. We will send qubits through quantum computer circuits and compute the outcome using matrices.

Using our tools to represent circuits with matrices, we will discover the kinds of operations that a circuit can perform. We'll combine these operations using matrix arithmetic. To bring these concepts into crystal clear focus, we'll write code to run on a quantum computer.

We'll cover the following topics in this chapter:

- Matrices for qubit states and operations
- Reversible operations
- Rotating the Bloch sphere around an axis

Matrices for qubit states and operations

In *Chapter 1, New Ways to Think about Bits*, we represented bits with kets and vectors:

$$|0\rangle = \begin{pmatrix} 1 \\ 0 \end{pmatrix} \qquad |1\rangle = \begin{pmatrix} 0 \\ 1 \end{pmatrix}$$

Let's expand that notation for qubits.

When you apply a Hadamard (h) gate to a $|0\rangle$ qubit, the qubit goes into a halfway state. Here's how we represent that halfway state in **Dirac** notation:

$$\frac{1}{\sqrt{2}}|0\rangle \;+\; \frac{1}{\sqrt{2}}|1\rangle \;=\; \frac{1}{\sqrt{2}}(\,|0\rangle \;+\; |1\rangle\,)$$

And here's how we represent that state with a vector:

$$\begin{pmatrix} \dfrac{1}{\sqrt{2}} \\[2ex] \dfrac{1}{\sqrt{2}} \end{pmatrix} = \frac{1}{\sqrt{2}}\begin{pmatrix} 1 \\ 1 \end{pmatrix}$$

This qubit state crops up so often in quantum computing that it's convenient to give the state its own symbol. We put a plus sign inside a ket and say that $|+\rangle \;=\; \frac{1}{\sqrt{2}}(\,|0\rangle \;+\; |1\rangle\,)$.

In vector notation, the vector's top number represents the *amount of* $|0\rangle$*'ness*, while the bottom number represents an equal amount of $|1\rangle$*'ness*. But what do the square roots do? What follows is a slight simplification.

> **Important note**
>
> A qubit's state has two parts. When you take the square of either part, you get a probability.

The two numbers associated with a qubit's state are called **amplitudes**.

> **Important note**
>
> So, we rephrase the previous statement as follows: "A qubit's state has two parts. Each of these parts is called an **amplitude**. When you square the absolute value of either amplitude, you get a probability."

Here's how you find the probability of getting 0 or 1 when you measure a qubit that's in the $|+\rangle$ state:

$$\text{Probability of getting 0} = \left|\frac{1}{\sqrt{2}}\right|^2 = \frac{1}{2}$$

$$\text{Probability of getting 1} = \left|\frac{1}{\sqrt{2}}\right|^2 = \frac{1}{2}$$

$$\text{Probability of getting either 0 or 1} = \frac{1}{2} + \frac{1}{2} = 1$$

To find out why the absolute value signs are needed, see *Chapter 9*.

Notice that **probabilities** are values such as $\frac{1}{2}$ and 1. Here are some examples:

- An event with probability 0 never takes place

- An event with probability 1 always takes place

- An event with probability $\frac{1}{2}$ takes place roughly 500 out of 1,000 times, roughly 2,500 out of 5,000 times, or roughly 5,000 out of 10,000 times, and so on

Probabilities and proportions are closely related. Imagine a qubit in the $|+\rangle$ state. The probability of getting 0 when you measure it is $\frac{1}{2}$, which is the same as 0.500. But when you measure 1,000 of these qubits, the proportion of 0s that you measure might be 0.496 or any other number close to 0.500.

Occasionally, a percentage sounds better than a fraction, so I write *50 percent probability* when I really mean *probability* $\frac{1}{2}$. But scientists generally represent probabilities with numbers from 0 to 1.

Qubits on the Bloch sphere

In *Chapter 2, What is a Qubit?*, we introduced the Bloch sphere as a way of visualizing qubit states. We identified the north pole with $|0\rangle$, the south pole with $|1\rangle$, and *halfway* as a point on the equator. With this chapter's matrix notation, we can provide more details about the sphere's values.

On the Bloch sphere, an **axis** is any line going from the center of the sphere to the surface of the sphere. The Bloch sphere has three special axes labeled X, Y, and Z (see *Figure 3.1*):

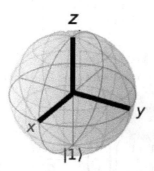

Figure 3.1 – The Bloch sphere's X, Y, and Z axes

When Qiskit draws the sphere, you normally don't see a label for the Z-axis. Instead, Qiskit puts $|0\rangle$ at the top of the sphere. Qiskit is reminding you that a qubit at the top of the sphere is in the $|0\rangle$ state. In the previous section, we gave $\frac{1}{\sqrt{2}}(|0\rangle + |1\rangle)$ the name $|+\rangle$. A qubit in the $|+\rangle$ state lies on a particular point along the equator. It's where the positive X-axis meets the equator, as shown in *Figure 3.2*:

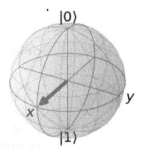

Figure 3.2 – The $\frac{1}{\sqrt{2}}$(|0⟩ + |1⟩) qubit

In *Chapter 2, What is a Qubit?*, we hinted that a Hadamard gate can put a qubit into the |+⟩ state. To find out why, let's look at the matrix representation of the Hadamard gate:

$$H = \begin{pmatrix} \dfrac{1}{\sqrt{2}} & \dfrac{1}{\sqrt{2}} \\ \dfrac{1}{\sqrt{2}} & -\dfrac{1}{\sqrt{2}} \end{pmatrix} = \frac{1}{\sqrt{2}}\begin{pmatrix} 1 & 1 \\ 1 & -1 \end{pmatrix}$$

When you multiply the Hadamard matrix by the |0⟩ qubit, you get the following:

$$H\,|0⟩ \; = \; \frac{1}{\sqrt{2}}\begin{pmatrix} 1 & 1 \\ 1 & -1 \end{pmatrix}\begin{pmatrix} 1 \\ 0 \end{pmatrix} = \frac{1}{\sqrt{2}}\begin{pmatrix} 1 \\ 1 \end{pmatrix} = \begin{pmatrix} \dfrac{1}{\sqrt{2}} \\ \dfrac{1}{\sqrt{2}} \end{pmatrix} = |+⟩$$

What happens when you start with |1⟩ and apply the Hadamard gate? In that situation, you get something slightly different:

$$H\,|1⟩ \; = \; \frac{1}{\sqrt{2}}\begin{pmatrix} 1 & 1 \\ 1 & -1 \end{pmatrix}\begin{pmatrix} 0 \\ 1 \end{pmatrix} = \frac{1}{\sqrt{2}}\begin{pmatrix} 1 \\ -1 \end{pmatrix} = \begin{pmatrix} \dfrac{1}{\sqrt{2}} \\ -\dfrac{1}{\sqrt{2}} \end{pmatrix}$$

The minus sign in the vector's lower entry puts this qubit on the opposite side of the Bloch sphere. It's where the *negative X*-axis meets the equator (see *Figure 3.3*):

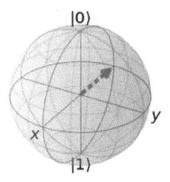

Figure 3.3 – The $\frac{1}{\sqrt{2}}$($|0\rangle$ $-$ $|1\rangle$) qubit

To represent this state, we put a minus sign inside a ket as follows: $|-\rangle$ $= \frac{1}{\sqrt{2}}$($|0\rangle$ $-$ $|1\rangle$).

> **Important note**
>
> In this book, most images of the Bloch sphere are screenshots of Qiskit code output. In some cases, we modify these screenshots to emphasize some aspect of the image. For example, in *Figure 3.3*, we've turned the arrow's solid line into a dotted line. This helps you see at a glance that the arrow points from the center of the sphere to the back of the sphere.

Since the $|-\rangle$ qubit is on the sphere's equator, we see roughly the same number of 0s and 1s when we measure many of these qubits. The math bears this out:

$$\text{Probability of getting 0} = \left| \frac{1}{\sqrt{2}} \right|^2 = \frac{1}{2}$$

$$\text{Probability of getting 1} = \left| -\frac{1}{\sqrt{2}} \right|^2 = \frac{1}{2}$$

$$\text{Probability of getting either 0 or 1} = \frac{1}{2} + \frac{1}{2} = 1$$

If you compare this calculation with the similar one for $|+\rangle$, you might be wondering why we bother keeping track of the minus sign in $-\frac{1}{\sqrt{2}}|1\rangle$. If so, read on.

In addition to its probabilities, a qubit has a **phase**. Think of a qubit as a pair of waves. The $|+\rangle$ qubit's waves look like the ones in *Figure 3.4*:

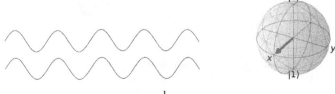

Figure 3.4 – The $\frac{1}{\sqrt{2}}$($|0\rangle$ $+$ $|1\rangle$) qubit

And the $|-\rangle$ qubit's waves look like the ones in *Figure 3.5*:

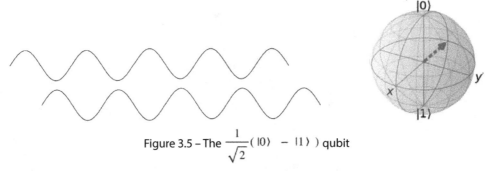

Figure 3.5 – The $\dfrac{1}{\sqrt{2}}(\ |0\rangle\ -\ |1\rangle\ )$ qubit

In *Figure 3.4*, with a plus sign between $|0\rangle$ and $|1\rangle$, the top and bottom waves are synchronized. But with the minus sign in *Figure 3.5*, the top and bottom waves are out of sync. In *Figure 3.5*, the top wave's trough meets the bottom wave's crest. You can imagine a transition from *Figure 3.4* to *Figure 3.5* in which the bottom wave slides rightward.

As a thought experiment, let's ask what happens if we put a second negative sign between $|0\rangle$ and $|1\rangle$ in the caption of *Figure 3.5*. Two negative signs make one positive sign. So, we're back to the $|+\rangle$ qubit of *Figure 3.4*. And, sure enough, when you start with *Figure 3.5*, and you slide the bottom wave rightward a second time, you get back to the synchronized waves in *Figure 3.4*.

In the same way, you can imagine the Bloch sphere's arrow moving halfway around the equator as it goes from the diagram in *Figure 3.4* to the one in *Figure 3.5*. An additional minus sign in the caption of *Figure 3.5* would move the arrow halfway around again, bringing it back to the sphere in *Figure 3.4*.

A qubit's phase is important because qubits with opposite phases can cancel each other out. For more details, see this chapter's *Reversing a matrix operation* section.

More points on the Bloch sphere

Going from *Figure 3.4* to *Figure 3.5*, you move halfway around the Bloch sphere's equator. To move by a lesser amount, you use complex numbers – numbers involving i, the square root of -1. *Figure 3.6* shows the phase and Bloch sphere for the qubit with state $\dfrac{1}{\sqrt{2}}|0\rangle\ +\ i\cdot\dfrac{1}{\sqrt{2}}|1\rangle$:

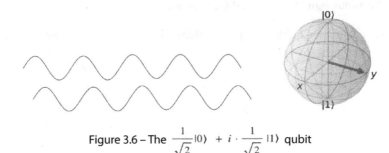

Figure 3.6 – The $\dfrac{1}{\sqrt{2}}|0\rangle\ +\ i\cdot\dfrac{1}{\sqrt{2}}|1\rangle$ qubit

In *Figure 3.6*, the imaginary number, i, pushes the Bloch sphere's arrow one-quarter of the way around the equator. In the wave diagram, it pushes the bottom wave one-quarter of the way toward resynchronizing.

Figure 3.7 shows the phase and Bloch sphere for $\frac{1}{\sqrt{2}}|0\rangle - i \cdot \frac{1}{\sqrt{2}}|1\rangle$:

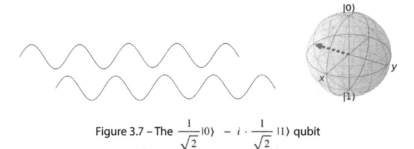

Figure 3.7 – The $\frac{1}{\sqrt{2}}|0\rangle - i \cdot \frac{1}{\sqrt{2}}|1\rangle$ qubit

With the $- i$ in *Figure 3.7*, the bottom wave moves three-quarters of the way from synchronized to resynchronized and pushes the Bloch sphere's arrow three-quarters of the way around the equator.

For more on complex numbers and their relevance to qubit states, see *Chapter 9, Shor's Algorithm*.

The X gate

In *Chapter 1, New Ways to Think about Bits*, we defined the classical **NOT gate** with the $\begin{pmatrix} 0 & 1 \\ 1 & 0 \end{pmatrix}$ matrix. The same matrix performs a NOT operation in quantum computing. It's called a NOT gate because, just as in classical computing, this matrix flips the amplitudes of $|0\rangle$ and $|1\rangle$:

$$\begin{pmatrix} 0 & 1 \\ 1 & 0 \end{pmatrix} \begin{pmatrix} 1 \\ 0 \end{pmatrix} = \begin{pmatrix} 0 \\ 1 \end{pmatrix}$$

$$\begin{pmatrix} 0 & 1 \\ 1 & 0 \end{pmatrix} \begin{pmatrix} \dfrac{1}{\sqrt{2}} \\ -\dfrac{1}{\sqrt{2}} \end{pmatrix} = \begin{pmatrix} -\dfrac{1}{\sqrt{2}} \\ \dfrac{1}{\sqrt{2}} \end{pmatrix}$$

When we rewrite these equations in Dirac notation, we get the following:

$$X \,|0\rangle$$

$$= X(\, 1\,|0\rangle \; + \; 0\,|1\rangle\,)$$

$$= \; (\, 0\,|0\rangle \; + \; 1\,|1\rangle\,)$$

$$= \,|1\rangle$$

$$X\left(\frac{1}{\sqrt{2}}\,|0\rangle \; - \; \frac{1}{\sqrt{2}}\,|1\rangle \right)$$

$$- \frac{1}{\sqrt{2}}\,|0\rangle \; + \; \frac{1}{\sqrt{2}}\,|1\rangle$$

Another name for the NOT gate is the **X gate** because it rotates the Bloch sphere 180 degrees around its *X*-axis:

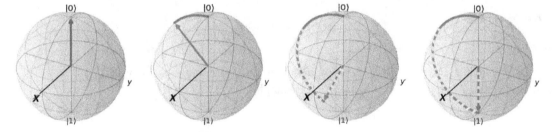

Figure 3.8 – Rotation around the X-axis

In *Figure 3.8*, the qubit starts at the sphere's |0⟩ north pole, moves downward along a longitudinal line, passes the equator, and finishes its journey at the |1⟩ south pole. During this rotation, the *X*-axis never moves. Once again, we've modified the drawings Qiskit makes by turning some solid lines into dotted lines. You may notice that, in *Figure 3.8*, the south pole's arrow is a dotted line. We made this a dotted line because, in Qiskit's drawings, the sphere is tilted forward. Look carefully, and you'll see that the north pole faces us and the south pole is on the back of the sphere. In most other figures, we've chosen not to turn this south pole arrow into a dotted line.

In the next section, we'll rotate around a different line on the sphere.

The Hadamard rotation

Every quantum gate performs a rotation of some kind. Take, for example, your friend –the Hadamard gate. Imagine a line from the center of the Bloch sphere to a point that's only $\frac{1}{4}$ of the way between $|0\rangle$ and $|1\rangle$. For want of a better name, let's assign the $\searrow$ symbol to this line (see *Figure 3.9*):

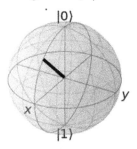

Figure 3.9 – The $\searrow$ line

If you start with a qubit in state$|0\rangle$ and rotate the sphere around this $\searrow$ line, you'll get the $\frac{1}{\sqrt{2}}(\,|0\rangle\ +\ |1\rangle\,)$ state. *Figure 3.10* shows this rotation in action:

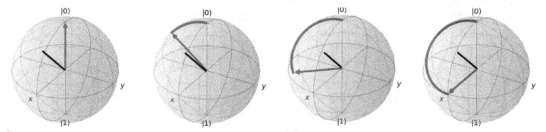

Figure 3.10 – Rotating the sphere around the $\searrow$ line

Earlier in this chapter, we showed that the Hadamard gate turns a $|0\rangle$ qubit into a $\frac{1}{\sqrt{2}}(\,|0\rangle\ +\ |1\rangle\,)$ qubit. So, it's reasonable to guess that applying the Hadamard gate always rotates the sphere around this $\searrow$ line. That is certainly what happens. Recall from the discussion surrounding *Figure 3.3* that the Hadamard gate moves a qubit from $|1\rangle$ (the sphere's south pole) to $|-\rangle$ (the equator on the far side). Take a moment to imagine what happens to the $|1\rangle$ point as you rotate the entire sphere around the $\searrow$ line. It's true! The $|1\rangle$ point moves to the far side's equator.

So far, we've represented quantum phenomena as matrices, wave pairs, and points on the Bloch sphere. In the next section, we'll implement these concepts as Qiskit code.

Combining gates along a single wire

A typical quantum computing circuit has several gates. With Qiskit code, we can explore the effects of applying one gate after another. To learn all about it, follow these steps:

1. In a Qiskit notebook, define the following circuit:

```
from qiskit import QuantumRegister, QuantumCircuit
reg = QuantumRegister(1)
circuit = QuantumCircuit(reg)
circuit.x(reg[0])
circuit.h(reg[0])
display(circuit.draw('latex', initial_state=True))
```

When you run this code, you'll get the output shown in *Figure 3.11*:

$$q0 : |0\rangle \ \boxed{\text{X}}\ \boxed{\text{H}}$$

Figure 3.11 – An X gate followed by a Hadamard gate

In the code, the call to draw with its initial_state=True parameter tells Qiskit to display |0⟩ at the beginning of the qubit wire. The display serves as a friendly reminder. It doesn't affect the way the circuit works.

Notice that this circuit has no measurement gate. We want Qiskit to draw a Bloch sphere of this circuit's output. But, if a circuit has a measurement gate, Qiskit doesn't let you draw a Bloch sphere. This makes sense because a measurement destroys a qubit's quantum state. Once you've measured a qubit, its state is either 0 or 1—nothing else.

2. To draw a Bloch sphere of the circuit's output, run this code:

```
from qiskit.quantum_info import Statevector
from qiskit.visualization import plot_bloch_multivector
vector = Statevector(circuit)
display(plot_bloch_multivector(vector.data))
```

When you run this code, Qiskit will display the sphere shown in *Figure 3.12*:

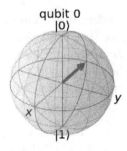

Figure 3.12 – Qiskit displays the |–⟩ qubit

This circuit's qubit begins its life in the |0⟩ state. The X gate puts that qubit into the |1⟩ state, and then the Hadamard gate puts the qubit into the |–⟩ state.

Take a moment to compare *Figure 3.3* and *Figure 3.12*. Both figures represent the same |−⟩ state, but I've modified *Figure 3.3* to make it clear that the arrow points toward the back of the sphere. In contrast, *Figure 3.12* is exactly what you get when you run the Qiskit code.

To display the sphere in *Figure 3.12*, we first call `Statevector(circuit)`. This call determines the state of the qubit after it passes through all the circuit's gates.

The `vector` variable contains various pieces of information about that state. Here are some examples:

- A call to `print(vector.num_qubits)` would display the number 1 because the circuit in *Figure 3.11* has only one qubit.

- A call to `print(vector.dim)` would display the number 2 because the circuit's one and only qubit state has two amplitudes — a |0⟩ amplitude and a |1⟩ amplitude.

- The `vector.data` expression contains the two amplitudes ($\frac{1}{\sqrt{2}}$ and $-\frac{1}{\sqrt{2}}$). The call to `plot_bloch_multivector` displays these amplitudes as they appear on a Bloch sphere.

> **Important note**
>
> Calling the `Statevector` function is a form of cheating. Any attempt to measure a qubit's state yields either 0 or 1. When you call `Statevector`, you're not measuring anything. You're just asking Qiskit to tell you what can be deduced by analyzing the circuit's gates. You could come to the same conclusion with matrix arithmetic or by reasoning about the Bloch sphere.

3. Let's add a measurement gate and run the circuit that we created in *Step 1*:

```
from qiskit import Aer, execute
device = Aer.get_backend('qasm_simulator')

circuit.measure_all()
job = execute(circuit, backend=device, shots=1000)
print(job.job_id())

result = job.result()
counts = result.get_counts(circuit)
print(counts)
```

The output of the run will look something like this:

$$\{'0': 505, '1': 495\}$$

When you run this code, the counts of 0s and 1s might be slightly different. But, chances are, the counts in your run are both close to 500. One way or another, the two counts always add up to 1,000 (the number of `shots` in the call to `execute`).

In *Figure 3.13*, we're using matrices to verify that the run's output makes sense:

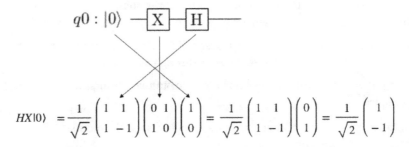

Figure 3.13 – Calculating the effect of applying X followed by H

In *Figure 3.13*, notice how you list mathematical operators in an order that's opposite to the way they appear in the circuit diagram. If you're not careful about this, you can get the wrong answer. (Take my word for it. When you do it in front of a class of college students, it's really embarrassing!)

4. Add one line to the code in *Step 1* and run all three cells again (by pressing *Shift + Enter* in each cell):

```
circuit.x(reg[0])
circuit.h(reg[0])
circuit.x(reg[0])
```

Figure 3.14 shows the output from the first two cells. (The arrow points toward the back of the Bloch sphere.)

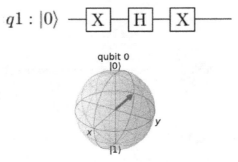

Figure 3.14 – Three gates in a row

Compare this Bloch sphere with the output in *Step 2* (*Figure 3.12*). Apparently, this step's additional X gate has no effect. You can think of this in more than one way:

*   After applying X and then Hadamard, the qubit's arrow points along the x-axis (from the middle to the back of the sphere). Imagine rotating the entire sphere around the X-axis, as in *Figure 3.8*. When you do this, the arrow in *Figure 3.14* doesn't move.

- Do the matrix multiplication:

$$XHX|0\rangle = \begin{pmatrix} 0 & 1 \\ 1 & 0 \end{pmatrix} \frac{1}{\sqrt{2}} \begin{pmatrix} 1 & 1 \\ 1 & -1 \end{pmatrix} \begin{pmatrix} 0 & 1 \\ 1 & 0 \end{pmatrix} \begin{pmatrix} 1 \\ 0 \end{pmatrix} = \begin{pmatrix} 0 & 1 \\ 1 & 0 \end{pmatrix} \frac{1}{\sqrt{2}} \begin{pmatrix} 1 \\ -1 \end{pmatrix} = \frac{1}{\sqrt{2}} \begin{pmatrix} -1 \\ 1 \end{pmatrix}$$

- Instead of $\frac{1}{\sqrt{2}} \begin{pmatrix} 1 \\ -1 \end{pmatrix}$, you end up with $\frac{1}{\sqrt{2}} \begin{pmatrix} -1 \\ 1 \end{pmatrix}$. But, as it turns out, having the minus sign on top instead of on the bottom makes no difference. The values on the top and bottom of the vector don't matter so much as the relationship between those values. When one value is the negative of the other value, the qubit goes to the back of the sphere.

5. Replace the three statements at the start of *Step 4* with the following two statements:

```
circuit.h(reg[0])
circuit.h(reg[0])
```

We'll run all three cells again (by pressing *Shift + Enter* in each cell).

Figure 3.15 shows the output from the three Qiskit cells:

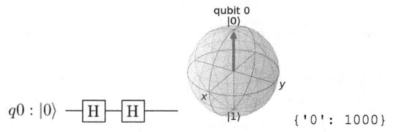

Figure 3.15 – Applying two Hadamard gates in a row

Applying the Hadamard gate twice in a row is like doing nothing. If you start with |0⟩, you end up with |0⟩. If you start with |1⟩, you end up with |1⟩. If you start with |+⟩, you end up with |+⟩, and so on. To find out why, see this chapter's *Reversing a matrix operation* section.

6. Now, from *Step 3*, replace device = Aer.get_backend('qasm_simulator') with this:

```
from qiskit_ibm_provider import IBMProvider
provider = IBMProvider()
device = provider.get_backend('ibmq_lima')
```

If ibmq_lima isn't available, find another backend that doesn't have simulator in its name. (For details, refer to *the Finding a backend* section of *Chapter 2, What is a Qubit?*)

When you press *Shift + Enter*, you're running the *HH* circuit on a real quantum computer. You may have to wait a while for the output. (I've seen jobs that take three minutes and jobs that take three days.). When you finally get the output, it may look something like this:

$$\{ \texttt{'0'}:\ 970,\ \texttt{'1'}:\ 30 \}$$

> **Tip**
>
> If you hover over the icons along the left edge of the Jupyter notebook, you'll find an icon whose hover tip is Jobs. When you click that icon, you'll see a list of jobs. You can click on a job whose status is Pending and get more information about that job. If the job is taking too long, and you want to cancel it, look for the Options button.

In *Step 5*, you got 1,000 zeros. But that's because you were running on a classical computer that was simulating a quantum computer. When you run on real quantum hardware, your results may vary. Here, in the 2020s decade, real quantum computers are noisy. You don't always get the answers you expect. That's because it's very difficult to keep qubits in their desired states.

Qubits suffer from **decoherence** – the natural loss of information due to inevitable interactions with the environment. As qubits travel from one point to another, they interact with any particles that they encounter. Like the measurement gates in our quantum circuits, these interactions change the qubits' states:

- When a qubit's state lies on the surface of the Bloch sphere, the qubit is in a **pure state**.

  In this book, the word "*state*" is shorthand for the phrase "*pure state*." Each of our qubits lives on the surface of the sphere.

- When a qubit's state lies in the Bloch sphere's interior, the qubit is in a **mixed state**.

  A mixed state is a combination of pure states weighted by each of the pure states' probabilities. If that doesn't mean much to you, don't worry. All you need to know is that decoherence leads to mixed states, and mixed states make our quantum computations inaccurate.

Engineers have learned to limit some of the decoherence in quantum computers, but not nearly enough to make quantum computing commercially feasible.

In this section, we learned that we can represent a quantum gate with a matrix. Does it work the other way? Does every matrix represent a quantum gate? To learn the answer, read the next section.

Reversible operations

You're composing a letter to the love of your life. You accidentally delete an entire paragraph. So, what do you do? Do you try to recreate the paragraph word by word? No. You use the document editor's Undo feature to go back to the time when the paragraph appeared.

Some things in life can be undone, whereas others can't. If you shoot a missile into the air, you can't get it back onto the launch pad unscathed. If the love of your life reads your letter, you can't *un-write* what you wrote. If someone sends bits through a classical **AND gate** and you examine the gate's output, you can't always tell what the input bits were. *Figure 3.16* illustrates this point.

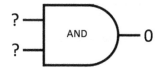

Figure 3.16 – Are the inputs 00, 01, or 10?

What about this chapter's quantum gates? Can they be undone? Pick any starting point on the Bloch sphere and then apply the X gate. When you apply the X gate, you rotate the sphere around the X-axis. *Figure 3.17*, shows what happens when you rotate $\frac{1}{\sqrt{2}}|0\rangle + i \cdot \frac{1}{\sqrt{2}}|1\rangle$ around the X-axis:

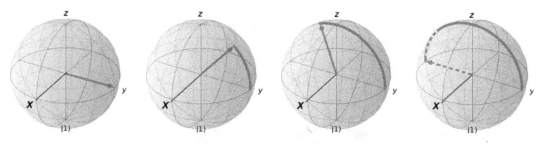

Figure 3.17 – Applying the X gate to $\frac{1}{\sqrt{2}}|0\rangle + i \cdot \frac{1}{\sqrt{2}}|1\rangle$

Whenever you apply an operation to a qubit, the entire Bloch sphere moves the same way all at once. The points on the sphere follow one another as if they're dancing together. If someone shows you the rightmost sphere in *Figure 3.17* and tells you that the dotted arrow got there after applying the X gate, can you figure out where the arrow started? You can! Simply rotate the sphere backward around the X-axis to get to the leftmost sphere in *Figure 3.17*. In the language of quantum computing, the application of an X gate is **reversible**.

Let's distinguish between what I call *genuine* quantum gates and *impostor* quantum gates. Almost all the gates that I will and have described in this book are genuine. The only impostor worth mentioning now is the gate that performs a measurement. When you measure a qubit, you always get either 0 or 1. After performing the measurement, you can't tell whether the qubit started in the $|0\rangle$ state, the $\frac{1}{\sqrt{2}}|0\rangle + i \cdot \frac{1}{\sqrt{2}}|1\rangle$ state, or some other exotic quantum state. The measurement operation isn't reversible.

Applying a measurement gate cannot be undone. But that's okay because the measurement gate is a kind of imposter among all the quantum gates. If you want reversibility, stick with the genuine gates.

> **Important note**
> Every genuine quantum operation is reversible.

The rotation in *Figure 3.17* lends evidence to this reversibility claim.

The next section will show what reversibility means for matrix multiplication.

Reversing a matrix operation

Consider the 2 x 2 matrix, $\begin{pmatrix} 1 & 0 \\ 0 & 1 \end{pmatrix}$, and the 3 x 3 matrix, $\begin{pmatrix} 1 & 0 & 0 \\ 0 & 1 & 0 \\ 0 & 0 & 1 \end{pmatrix}$.

Both of these matrices have 1s along the diagonal and 0s everywhere else. A matrix with 1s along the diagonal and 0s everywhere else is called an **identity matrix** (normally denoted by the uppercase letter *I*). You can form identity matrices of any size, such as 4 x 4, 100 x 100, and so on.

Multiplying a matrix by an identity matrix on either side doesn't change anything – that is, for any matrix, *M*, we get the following:

$$M\,I = M$$

$$I\,M = M$$

Try it yourself: Multiply $\begin{pmatrix} 2 & 3 \\ -7 & 4 \end{pmatrix}$ by $\begin{pmatrix} 1 & 0 \\ 0 & 1 \end{pmatrix}$. The answer you'll get is $\begin{pmatrix} 2 & 3 \\ -7 & 4 \end{pmatrix}$.

You can liken an identity matrix to scalar number 1:

$$5 \cdot 1 = 1 \cdot 5 = 5$$

Consider these calculations:

$$\begin{pmatrix} 2 & 5 \\ 4 & 3 \end{pmatrix} \begin{pmatrix} -\dfrac{3}{14} & \dfrac{5}{14} \\ \dfrac{4}{14} & -\dfrac{2}{14} \end{pmatrix} = \begin{pmatrix} -\dfrac{6}{14}+\dfrac{20}{14} & \dfrac{10}{14}-\dfrac{10}{14} \\ -\dfrac{12}{14}+\dfrac{12}{14} & \dfrac{20}{14}-\dfrac{6}{14} \end{pmatrix} = \begin{pmatrix} 1 & 0 \\ 0 & 1 \end{pmatrix} = I$$

$$\begin{pmatrix} -\dfrac{3}{14} & \dfrac{5}{14} \\ \dfrac{4}{14} & -\dfrac{2}{14} \end{pmatrix} \begin{pmatrix} 2 & 5 \\ 4 & 3 \end{pmatrix} = \begin{pmatrix} -\dfrac{6}{14}+\dfrac{20}{14} & -\dfrac{15}{14}+\dfrac{15}{14} \\ \dfrac{8}{14}-\dfrac{8}{14} & \dfrac{20}{14}-\dfrac{6}{14} \end{pmatrix} = \begin{pmatrix} 1 & 0 \\ 0 & 1 \end{pmatrix} = I$$

When you multiply $\begin{pmatrix} 2 & 5 \\ 4 & 3 \end{pmatrix}$ on either side by $\begin{pmatrix} -\dfrac{3}{14} & \dfrac{5}{14} \\ \dfrac{4}{14} & -\dfrac{2}{14} \end{pmatrix}$, you get the identity matrix. Because of

this, we call $\begin{pmatrix} -\dfrac{3}{14} & \dfrac{5}{14} \\ \dfrac{4}{14} & -\dfrac{2}{14} \end{pmatrix}$ the **inverse** of $\begin{pmatrix} 2 & 5 \\ 4 & 3 \end{pmatrix}$.

You can liken a matrix's inverse to a scalar number's reciprocal:

$$5 \cdot \frac{1}{5} = \frac{1}{5} \cdot 5 = 1$$

We can call $\frac{1}{5}$ the *inverse* of 5.

To denote the inverse of a matrix, we use a superscript of -1. So, for example, if M is $\begin{pmatrix} 2 & 5 \\ 4 & 3 \end{pmatrix}$, then M^{-1} is $\begin{pmatrix} -\frac{3}{14} & \frac{5}{14} \\ \frac{4}{14} & -\frac{2}{14} \end{pmatrix}$. To express the fact that a matrix multiplied by its inverse is the identity, we can write the following:

$$M\left(M^{-1}\right) = I$$
$$\left(M^{-1}\right)M = I$$

Important note

Matrix multiplication isn't commutative. You can't always count on MN being equal to NM. But, when you multiply a matrix by its inverse or either side, you always get I. To find out if N is the inverse of M, you can check that $MN = I$. If that's true, you don't have to check that $NM = I$.

Some matrices don't have inverses. For example, you can't multiply $\begin{pmatrix} 3 & 2 \\ 6 & 4 \end{pmatrix}$ by another matrix to get the identity matrix. We say that the $\begin{pmatrix} 3 & 2 \\ 6 & 4 \end{pmatrix}$ matrix is not **invertible**.

On the other side of the spectrum, some matrices are their own inverses. The Hadamard matrix, $\frac{1}{\sqrt{2}}\begin{pmatrix} 1 & 1 \\ 1 & -1 \end{pmatrix}$, is one shining example of this:

$$\frac{1}{\sqrt{2}}\begin{pmatrix} 1 & 1 \\ 1 & -1 \end{pmatrix}\frac{1}{\sqrt{2}}\begin{pmatrix} 1 & 1 \\ 1 & -1 \end{pmatrix} = \frac{1}{2}\begin{pmatrix} 1\cdot 1+1\cdot 1 & 1\cdot 1+1\cdot(-1) \\ 1\cdot 1+(-1)\cdot 1 & 1\cdot 1+(-1)\cdot(-1) \end{pmatrix} = \begin{pmatrix} 1 & 0 \\ 0 & 1 \end{pmatrix} = I$$

No matter what you start with, applying the Hadamard operator twice brings you right back to where you started.

In the *Qubits on the Bloch sphere* section, we promised to show you how qubits' phases can cancel each other out. The Hadamard gate's self-inverting behavior provides a good example of this. Remember the following:

$$H|0\rangle = \frac{|0\rangle + |1\rangle}{\sqrt{2}} \quad \text{and} \quad H|1\rangle = \frac{|0\rangle - |1\rangle}{\sqrt{2}}$$

Now, start with $|0\rangle$ and apply the H gate twice. Here's what you'll get:

$$HH|0\rangle = H\left(\frac{|0\rangle + |1\rangle}{\sqrt{2}}\right) = \frac{H|0\rangle + H|1\rangle}{\sqrt{2}} = \frac{\frac{|0\rangle + |1\rangle}{\sqrt{2}} + \frac{|0\rangle - |1\rangle}{\sqrt{2}}}{\sqrt{2}} = \frac{2|0\rangle}{2} = |0\rangle$$

The opposite-signed phases cancel one another out to give you the $|0\rangle$ state that you started with.

Unitary matrices

The **transpose** of a matrix is what you get when you flip a matrix around its backslash-like diagonal. The symbol for taking a transpose is an uppercase letter T superscript. Here are some examples:

$$\begin{pmatrix} 1 & 2 \\ 3 & 4 \end{pmatrix}^T = \begin{pmatrix} 1 & 3 \\ 2 & 4 \end{pmatrix} \qquad \begin{pmatrix} 1 & 2 & 3 & 4 \\ 5 & 6 & 7 & 8 \\ 9 & 10 & 11 & 12 \\ 13 & 14 & 15 & 16 \end{pmatrix}^T = \begin{pmatrix} 1 & 5 & 9 & 13 \\ 2 & 6 & 10 & 14 \\ 3 & 7 & 11 & 15 \\ 4 & 8 & 12 & 16 \end{pmatrix}$$

Every number in these matrices is a **real number** – a number that doesn't involve $\sqrt{-1}$. Real numbers include values such as 3, 0, -9, $\frac{473}{2156}$, 0.8273, 0.33333..., and π.

A number that has $\sqrt{-1}$ somewhere in its representation is an example of a **complex number**. This chapter doesn't do much with complex numbers, so we can make a slightly simplified assertion about matrices. A matrix with only real numbers represents a quantum computing operator if and only if the matrix is invertible and the inverse of the matrix is the matrix's transpose.

A matrix that contains only real numbers is called a **unitary matrix**, so long as it satisfies those two properties.

If we could make an action-packed movie about quantum computing, unitary matrices would be *the good guys*. When you read the words *unitary matrix*, you should think *genuinely reversible quantum gate*.

> **Important note**
>
> A matrix represents a quantum computing operator if and only if it's a unitary matrix.

When you transpose the Hadamard matrix, you don't change anything. You get that same Hadamard matrix back. And, sure enough, the Hadamard matrix is its own inverse. So, the Hadamard matrix is unitary. That's how we know that the Hadamard matrix represents a quantum operator.

The X gate's matrix is also unitary:

$$X(X^T) = (X^T)X = \begin{pmatrix} 0 & 1 \\ 1 & 0 \end{pmatrix}\begin{pmatrix} 0 & 1 \\ 1 & 0 \end{pmatrix} = \begin{pmatrix} 1 & 0 \\ 0 & 1 \end{pmatrix} = I$$

So, X is a unitary matrix and therefore represents a quantum computing operator.

So far in this chapter, we've examined rotations of the Bloch sphere and reversible quantum operations. We've used matrices and other tools to connect these two concepts. In the next section, we'll look carefully at some specific Bloch sphere rotations.

Rotating the Bloch sphere around an axis

Rotations of the Bloch sphere can move a qubit from any point on the sphere's surface to any other point on the surface. This section will show you some of Qiskit's most useful rotation commands.

Experimenting with rotations

Qiskit provides many functions that rotate the Bloch sphere and change a qubit's state. For hands-on practice with some of these rotations, open a new Qiskit notebook and follow these steps:

1. Run the following code:

```
from qiskit import QuantumRegister, QuantumCircuit
from math import pi
reg = QuantumRegister(1)
circuit = QuantumCircuit(reg)
circuit.ry(pi/2, reg[0])
display(circuit.draw('latex'))
```

Figure 3.18 shows you the circuit diagram that this code generates:

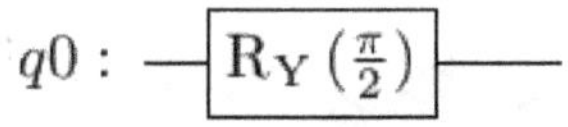

$$q0 : \boxed{\mathrm{R_Y}\left(\tfrac{\pi}{2}\right)}$$

Figure 3.18 – Rotating by $\dfrac{\pi}{2}$ radians about the Y-axis

The circuit's one and only qubit begins its life in the $|0\rangle$ state. Then, the `circuit.ry(pi/2, reg[0])` statement tells Qiskit to rotate the sphere $\dfrac{\pi}{2}$ radians around the Y-axis.

The value $\dfrac{\pi}{2}$ *radians* is the same as 90 degrees. If you're not familiar with radian units, read this chapter's *What is a radian?* section.

2. To draw the Bloch sphere, copy the code from *Step 2* of the *Combining gates along a single wire* section into your notebook's empty cell and press *Shift + Enter*.

In *Figure 3.19*, I've enhanced the code's output to convince you that `ry` performs a *Y*-axis rotation:

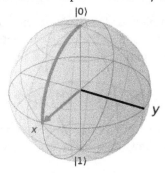

Figure 3.19 – Rotating by $\frac{\pi}{2}$ radians about the Y-axis on the Bloch sphere

3. Copy the code from *Step 3* of the *Combining gates along a single wire* section into this notebook's empty cell and run the code in that cell.

    The output you'll get will be something like `{'0': 501, '1': 499}`.

    So far in this section, we've dealt exclusively with qubits in the $|0\rangle$, $|1\rangle$, or halfway states. Quantum computing is, by its very nature, an analog process. The value that a qubit may assume isn't limited to a few discrete ticks. Instead, a qubit's state can range anywhere between $|0\rangle$ and $|1\rangle$, and taking phase into account adds even more possibilities. In the next step, we'll put a qubit into what I informally call a *one-quarter: three-quarters* state.

4. From the cell of *Step 1*, change `pi/2` to `pi/3` and run the code in the three cells again.

    What you might get is shown in *Figure 3.20*:

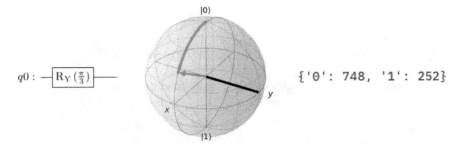

Figure 3.20 – Rotating by $\frac{\pi}{3}$ radians about the Y-axis

Once again, I've enhanced the drawing of the Bloch sphere.

This circuit's qubit starts in the $|0\rangle$ state, but the `circuit.ry(pi/3, reg[0])` statement tells Qiskit to rotate the sphere one-third of the way around the *Y*-axis.

The matrix representation of this circuit's action is as follows:

$$R_Y\left(\frac{\pi}{3}\right)|0\rangle = \begin{pmatrix} \dfrac{\sqrt{3}}{2} & -\dfrac{1}{2} \\ \dfrac{1}{2} & \dfrac{\sqrt{3}}{2} \end{pmatrix}\begin{pmatrix} 1 \\ 0 \end{pmatrix} = \begin{pmatrix} \dfrac{\sqrt{3}}{2} \\ \dfrac{1}{2} \end{pmatrix}$$

The resulting vector represents a qubit's state because the sum of the squares adds up to 1:

$$\left|\frac{\sqrt{3}}{2}\right|^2 + \left|\frac{1}{2}\right|^2 = \frac{3}{4} + \frac{1}{4} = 1$$

Notice the squares of the two amplitude values. When you measure this qubit, the probability of getting 0 is $\frac{3}{4}$, and the probability of getting 1 is $\frac{1}{4}$. That's why the output, $\{\,'0'\colon\ 749,\ '1'\colon\ 251\}$, makes sense. Qiskit counts roughly three times as many 0s as 1s.

Like every matrix that represents a genuine quantum computing operation, the matrix in this circuit is unitary. Here's how you can find out for sure:

$$\begin{pmatrix} \dfrac{\sqrt{3}}{2} & -\dfrac{1}{2} \\ \dfrac{1}{2} & \dfrac{\sqrt{3}}{2} \end{pmatrix}^T\begin{pmatrix} \dfrac{\sqrt{3}}{2} & -\dfrac{1}{2} \\ \dfrac{1}{2} & \dfrac{\sqrt{3}}{2} \end{pmatrix} = \begin{pmatrix} \dfrac{\sqrt{3}}{2} & \dfrac{1}{2} \\ -\dfrac{1}{2} & \dfrac{\sqrt{3}}{2} \end{pmatrix}\begin{pmatrix} \dfrac{\sqrt{3}}{2} & -\dfrac{1}{2} \\ \dfrac{1}{2} & \dfrac{\sqrt{3}}{2} \end{pmatrix} = \begin{pmatrix} 1 & 0 \\ 0 & 1 \end{pmatrix}$$

Where do the numbers in this matrix come from? To help answer this question, I have provided the general formula for a ry matrix:

$$R_Y(\theta) = \begin{pmatrix} \cos\dfrac{\theta}{2} & -\sin\dfrac{\theta}{2} \\ \sin\dfrac{\theta}{2} & \cos\dfrac{\theta}{2} \end{pmatrix}$$

Substituting $\dfrac{\pi}{3}$ for θ, you get the following:

$$R_Y\left(\frac{\pi}{3}\right) = \begin{pmatrix} \cos\dfrac{\pi}{6} & -\sin\dfrac{\pi}{6} \\ \sin\dfrac{\pi}{6} & \cos\dfrac{\pi}{6} \end{pmatrix} = \begin{pmatrix} \dfrac{\sqrt{3}}{2} & -\dfrac{1}{2} \\ \dfrac{1}{2} & \dfrac{\sqrt{3}}{2} \end{pmatrix}$$

You can look up the values of things such as $\cos\frac{\pi}{6}$ and $\sin\frac{\pi}{6}$ in many different places. But, if you want to understand how these values come about, read the remaining sections of this chapter.

What is a radian?

If we're serious about rotations, we have to work with angles. Most people know what *360 degrees* means. But, for many people, equating 360 degrees with 2π *radians* is unfamiliar territory.

When scientists describe angles, they generally don't use degrees. Instead, they use **radians**. To convert from degrees to radians, you multiply by $\frac{\pi}{180}$. To convert from radians to degrees, you multiply by $\frac{180}{\pi}$. Here are two examples:

$$60 \text{ degrees} = 60 \text{ degrees} \times \frac{\pi \text{ radians}}{180 \text{ degrees}} = \frac{60\pi \text{ radians}}{180} = \frac{\pi}{3} \text{ radians}$$

$$\frac{\pi}{2} \text{ radians} = \frac{\pi}{2} \text{ radians} \times \frac{180 \text{ degrees}}{\pi \text{ radians}} = \frac{180 \text{ degrees}}{2} = 90 \text{ degrees}$$

Notice how conversions of this kind work. Multiplying by $\frac{\pi \text{ radians}}{180 \text{ degrees}}$ is like multiplying by 1 because π radians = 180 degrees In the previous first calculation, the words *degrees* are canceled, so you're left with a number of radians. In the second calculation, the *radians* are canceled, so you end up with degrees.

A taste of trigonometry

In my experience of teaching university mathematics, nothing scares students more than trig functions. These functions change in strange, unintuitive ways and yield results that aren't easy to guess. At my university, we face the problem head-on by introducing trig functions early in the first math course. Does this strategy work? I don't know. The exams that we give indicate student success, but I don't trust exams of this kind.

Anyway, this section is a bare-bones introduction to trig functions. You can read this section now or come back to it if you ever need a trigonometry review.

First, imagine a right triangle seated on a graph. Consider the angle closest to where the graph's axes cross one another. The **sine** (denoted as sin) of that angle is the ratio of the triangle's height to its hypotenuse. You can see this in *Figure 3.21*.

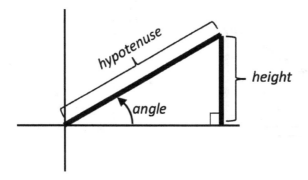

$$\sin angle = \frac{height}{hypotenuse}$$

Figure 3.21 – The sine of an angle

The **cosine** (denoted as cos) of that same angle is the ratio of the triangle's base to its hypotenuse. You can see this in *Figure 3.22*.

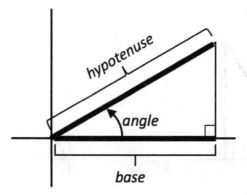

$$\cos angle = \frac{base}{hypotenuse}$$

Figure 3.22 – The cosine of an angle

With those facts in mind, *Figure 3.23* shows the ratios for a 30-degree angle (also known as a $\frac{\pi}{6}$ radian angle):

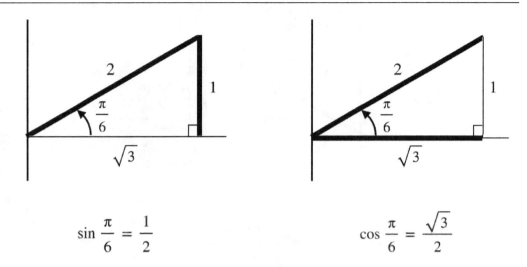

$$\sin \frac{\pi}{6} = \frac{1}{2} \qquad\qquad \cos \frac{\pi}{6} = \frac{\sqrt{3}}{2}$$

Figure 3.23 – Trig functions for $\frac{\pi}{6}$

With the same reasoning, you can determine the cosine and sine for some other angles. For example, a $\frac{\pi}{3}$ radian triangle is like a $\frac{\pi}{6}$ triangle tilted on its side. *Figure 3.24* illustrates this idea.

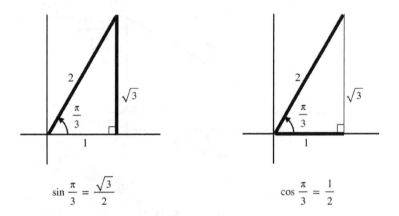

$$\sin \frac{\pi}{3} = \frac{\sqrt{3}}{2} \qquad\qquad \cos \frac{\pi}{3} = \frac{1}{2}$$

Figure 3.24 – Trig functions for $\frac{\pi}{3}$

A $\frac{\pi}{4}$ radian (45 degrees) triangle's base and height have equal sizes. You can see this in *Figure 3.25*.

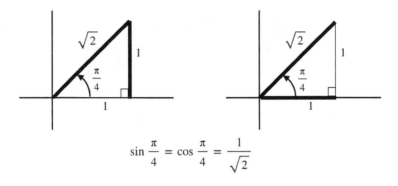

$$\sin \frac{\pi}{4} = \cos \frac{\pi}{4} = \frac{1}{\sqrt{2}}$$

Figure 3.25 – Trig functions for $\dfrac{\pi}{4}$

A triangle's base or height may be negative, but the hypotenuse is never negative. *Figure 3.26* shows you how this works.

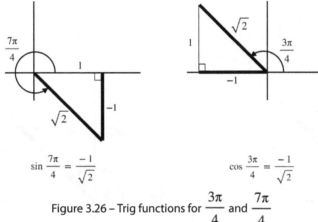

$$\sin \frac{7\pi}{4} = \frac{-1}{\sqrt{2}} \qquad\qquad \cos \frac{3\pi}{4} = \frac{-1}{\sqrt{2}}$$

Figure 3.26 – Trig functions for $\dfrac{3\pi}{4}$ and $\dfrac{7\pi}{4}$

Finally, when a triangle sits flush against an axis, either its base or its height is zero. You can see this in *Figure 3.27*.

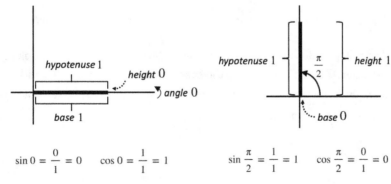

$$\sin 0 = \frac{0}{1} = 0 \qquad \cos 0 = \frac{1}{1} = 1 \qquad\qquad \sin \frac{\pi}{2} = \frac{1}{1} = 1 \qquad \cos \frac{\pi}{2} = \frac{0}{1} = 0$$

Figure 3.27 – Trig functions for 0 and $\dfrac{\pi}{2}$

Using these rules, you can find the sine and cosine for angles such as $\frac{\pi}{3}$, $\frac{5\pi}{6}$, π, $\frac{3\pi}{4}$, and many others.

The values repeat when you go around the circle more than once. So, for example, the trig functions for $\frac{5\pi}{4}$ are identical to those for $\frac{\pi}{4}$. *Figure 3.28* shows the two triangles.

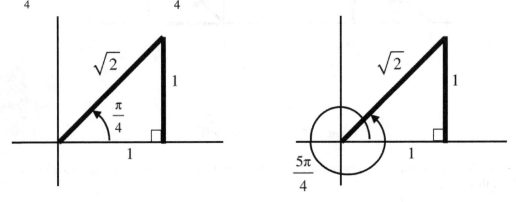

Figure 3.28 – The trig functions of $\frac{\pi}{4}$ and $\frac{5\pi}{4}$ have the same values

You can also have negative angles. You can see this in *Figure 3.29*.

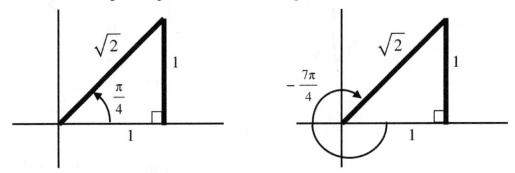

Figure 3.29 – The trig functions of $\frac{\pi}{4}$ and $-\frac{7\pi}{4}$ have the same values

Of course, many angles have ratios that aren't convenient for hand calculation. I can look up the ratios for $\frac{\pi}{9}$ and find that when the hypotenuse is 1, the base is approximately 0.9397, and the height is approximately 0.3402. But I have no reason to remember these numbers. When I need the sine and cosine of $\frac{\pi}{9}$, I just look them up on the web.

Summary

In this chapter, we developed the tools to describe a qubit. We changed a qubit's state by sending the qubit through a quantum gate. In mathematical terms, the gate applies an operator to the qubit – an operator that we represent using a matrix.

In classical computing, things start to become interesting when we execute statements conditionally. Most languages have `if` statements for conditional execution. To perform conditional execution at the quantum level, we need some kind of qubit-to-qubit interaction. One qubit changes if some other qubit's state is $|1\rangle$.

To achieve this interaction, we need some new operators. Each such operator acts on more than one qubit at a time. We use matrices to represent these operators. When we do, we discover some bizarre effects. That's what the next chapter is all about.

Questions

Answer the following questions to test your knowledge of this chapter:

1. Which of the following vectors represents a qubit state?

   A. $\dfrac{1}{\sqrt{7}} \begin{pmatrix} \sqrt{3} \\ -2 \end{pmatrix}$

   B. $\begin{pmatrix} 1 \\ 1 \end{pmatrix}$

   C. $\begin{pmatrix} 1 \\ 0 \\ 0 \end{pmatrix}$

2. In quantum computing, the Z gate rotates a Bloch sphere π radians around the Z-axis. Draw the result of applying a Z gate to a $|+\rangle$ qubit.

3. The matrix representation of a Z gate is $\begin{pmatrix} 1 & 0 \\ 0 & -1 \end{pmatrix}$. Check to make sure that this matrix is unitary.

4. Apply the Z gate matrix from *Question 3* to a $|+\rangle$ qubit. Does the result you get confirm your answer to *Question 2*?

5. Write Qiskit code to test the result you got in *Questions 2, 3,* and *4*.

6. The matrix representation of $R_Y\left(\dfrac{\pi}{2}\right)$ is $\dfrac{1}{\sqrt{2}} \begin{pmatrix} 1 & -1 \\ 1 & 1 \end{pmatrix}$. Check to make sure that this matrix is unitary.

7. Verify that the matrix representation of $R_Y\left(\dfrac{\pi}{2}\right)$ is $\dfrac{1}{\sqrt{2}} \begin{pmatrix} 1 & -1 \\ 1 & 1 \end{pmatrix}$. Use the last formula in this chapter's *Experimenting with rotations* section.

8. In *Step 2* of the *Experimenting with rotations* section, applying $R_Y\left(\dfrac{\pi}{2}\right)$ to $|0\rangle$ has the same effect as applying H to $|0\rangle$. Do the matrix calculation to show that $H|0\rangle = R_Y\left(\dfrac{\pi}{2}\right)|0\rangle$.

9. Draw pictures of the Bloch sphere to verify that $H|+\rangle \neq R_Y\left(\dfrac{\pi}{2}\right)|+\rangle$.

10. Use matrix multiplication to verify that $R_Y\left(\dfrac{\pi}{2}\right)$ is the same as XH.

Qubit Conspiracy Theories

Chapters 2 and *3* explained measuring individual qubits. You can send a qubit along a wire, measure the qubit, and add this measurement to your tally of results. That's fine for individual qubits. But in computing and life, things are more interesting when you start combining elements. So, in this chapter, we'll see what happens when one qubit interacts with another qubit.

We start by examining gates in which the state of one qubit depends on the state of another. It's like writing the following informal, English-language code:

```
if q0 is 0
    set q1 to 0
else
    set q1 to 1
```

Since the state of q0 can be somewhere between $|0\rangle$ and $|1\rangle$, the end result for q1 may be between $|0\rangle$ and $|1\rangle$. What's more, q0 and q1 may become mysteriously linked across time and space. They're like identical twins who are separated at birth. When one twin stubs her toe, the other twin says *Ouch*, no matter how many thousands of miles separate them from one another.

In 1964, physicist John Bell wrote a paper that paved the way for a deeper understanding of this strange phenomenon, and we'll cover that topic in some detail.

In this chapter, we'll cover the following topics:

- Multi-qubit gates
- Magic tricks with multi-qubit gates
- John Bell's amazing discovery
- Combining probabilities

Multi-qubit gates

In *Chapter 1, New Ways to Think about Bits*, the classical AND gate has two input bits but only one output bit. When you deal with qubits, this never happens. In quantum computing, each qubit's wire goes from the beginning to the very end of its circuit. This rule stems from the reversibility requirement that we described in *Chapter 3, Math for Qubits and Quantum Gates*.

With a quantum gate, the number of outputs must equal the number of inputs. When two qubits pass through such a gate, the two qubits are affected. This chapter explores gates that deal with two or more qubits. In this section, we'll describe several commonly used multi-qubit gates. Each of these gates is a building block in the construction of quantum computing algorithms.

CNOT and flipped CNOT gates

A **controlled NOT** (**CNOT**) gate involves two qubits. We call one qubit the **control qubit** and the other the **target qubit**. The control qubit controls whether the target qubit's state changes or not, as outlined here:

- When the control qubit is $|0\rangle$, neither qubit's value changes
- When the control qubit is $|1\rangle$, the computer applies NOT to the target qubit

The circuit diagrams in *Figure 4.1* illustrate the CNOT gate's rules:

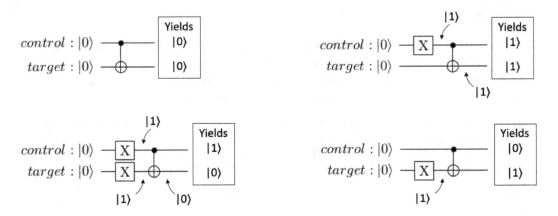

Figure 4.1 – CNOT gate

Figure 4.1 shows how you diagram a CNOT gate. The control qubit's wire gets a solid dot, and the target qubit's wire gets a circle with a plus sign inside it. A vertical line connects the two wires.

The controlled NOT's matrix representation is a 4-by-4 matrix:

$$\begin{pmatrix} 1 & 0 & 0 & 0 \\ 0 & 0 & 0 & 1 \\ 0 & 0 & 1 & 0 \\ 0 & 1 & 0 & 0 \end{pmatrix}$$

This matrix is its own transpose. Let's check to make sure that the matrix is unitary:

$$(CNOT^T)\,CNOT = \begin{pmatrix} 1 & 0 & 0 & 0 \\ 0 & 0 & 0 & 1 \\ 0 & 0 & 1 & 0 \\ 0 & 1 & 0 & 0 \end{pmatrix}\begin{pmatrix} 1 & 0 & 0 & 0 \\ 0 & 0 & 0 & 1 \\ 0 & 0 & 1 & 0 \\ 0 & 1 & 0 & 0 \end{pmatrix} = \begin{pmatrix} 1 & 0 & 0 & 0 \\ 0 & 1 & 0 & 0 \\ 0 & 0 & 1 & 0 \\ 0 & 0 & 0 & 1 \end{pmatrix} = I$$

Think about the upper-right circuit in *Figure 4.1*. How do you express this circuit as a collection of matrices? You start by taking the tensor product of the qubits in their initial states. *Figure 4.2* shows what you get:

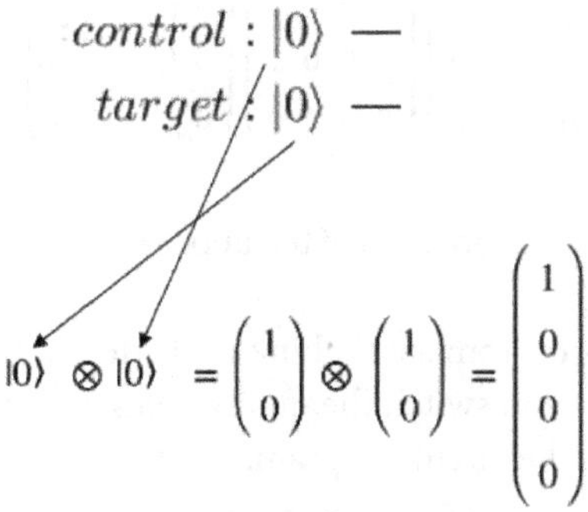

$$control : |0\rangle \ —$$
$$target : |0\rangle \ —$$

$$|0\rangle \otimes |0\rangle = \begin{pmatrix} 1 \\ 0 \end{pmatrix} \otimes \begin{pmatrix} 1 \\ 0 \end{pmatrix} = \begin{pmatrix} 1 \\ 0 \\ 0 \\ 0 \end{pmatrix}$$

Figure 4.2 – The |00⟩ qubits

In *Figure 4.3*, you combine the top qubit's X gate with the lack of a gate for the bottom qubit. The lack of a gate does nothing, so you represent it with an identity matrix:

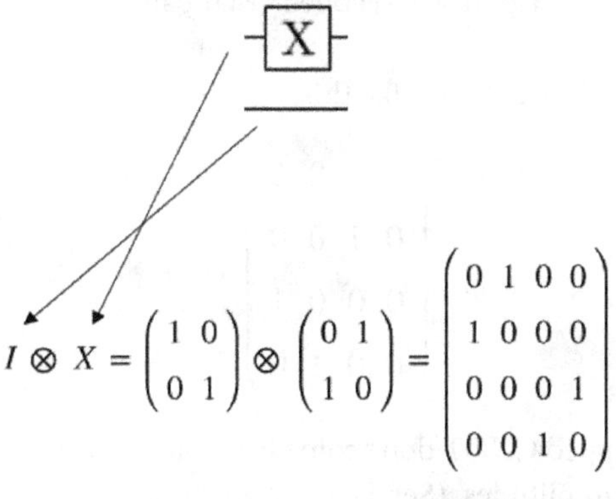

$$I \otimes X = \begin{pmatrix} 1 & 0 \\ 0 & 1 \end{pmatrix} \otimes \begin{pmatrix} 0 & 1 \\ 1 & 0 \end{pmatrix} = \begin{pmatrix} 0 & 1 & 0 & 0 \\ 1 & 0 & 0 & 0 \\ 0 & 0 & 0 & 1 \\ 0 & 0 & 1 & 0 \end{pmatrix}$$

Figure 4.3 – An X gate on the upper qubit

> **Important note**
>
> When you take the tensor product, the lower qubit's matrix is always on the left. If you ignore this rule, you may get an incorrect result.

As you move from left to right along the circuit, you don't use the tensor product. Instead, you combine the circuit's actions using ordinary matrix multiplication. The rightmost part of the circuit diagram belongs to the leftmost part of the calculation as in *Figure 4.4*:

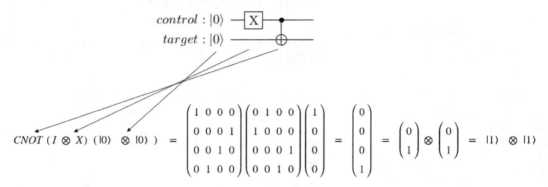

Figure 4.4 – Matrix representation of the upper-right circuit in Figure 4.1

When you apply a CNOT gate, it's customary to think of the top qubit as the control qubit and the bottom qubit as the target qubit. If you switch the qubits' roles, you have what's sometimes called a **flipped CNOT** gate. *Figure 4.5* has the circuit diagram.

$$target : —\oplus—$$
$$control : —\bullet—$$

Figure 4.5 – Flipped CNOT gate

The matrix for the flipped CNOT gate looks like this:

$$\begin{pmatrix} 1 & 0 & 0 & 0 \\ 0 & 1 & 0 & 0 \\ 0 & 0 & 0 & 1 \\ 0 & 0 & 1 & 0 \end{pmatrix}$$

The matrices for CNOT and flipped CNOT don't come from out of the blue. The CNOT matrix always swaps a vector's $|01\rangle$ and $|11\rangle$ amplitudes. (See *Figure 4.6*.)

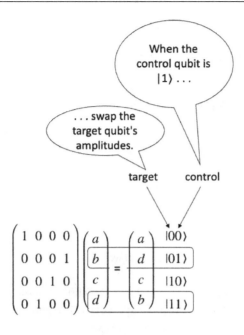

Figure 4.6 – What does the CNOT matrix do?

In a similar way, the flipped CNOT matrix always swaps a vector's $|10\rangle$ and $|11\rangle$ amplitudes. (See *Figure 4.7*.)

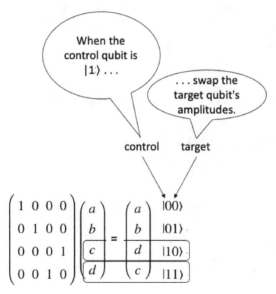

Figure 4.7 – What does the flipped CNOT matrix do?

Here's the Qiskit code to create a CNOT gate:

```
from qiskit import QuantumRegister, QuantumCircuit
reg = QuantumRegister(2)
circuit = QuantumCircuit(reg)
circuit.cnot(reg[0], reg[1])
display(circuit.draw('latex'))
```

The code's output is the circuit diagram shown in *Figure 4.8*:

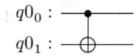

Figure 4.8 – Qiskit draws a CNOT gate

In the call to cnot, the first parameter (reg[0]) is the control qubit, and the second parameter (reg[1]) is the target qubit. That's how it always works. The general format for Qiskit's cnot function is shown here:

```
circuit.cnot(control_qubit, target_qubit)
```

Another name for Qiskit's cnot function is cx. You can use these names interchangeably.

Qiskit doesn't have a separate function for the flipped CNOT gate. Instead, a flipped CNOT gate is simply a cnot function with the parameters reversed. For example, in the code that accompanies *Figure 4.8*, you can modify the cnot call as follows:

```
circuit.cnot(reg[1], reg[0])
```

When you make this change, you get the output shown in *Figure 4.9*:

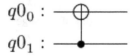

Figure 4.9 – Qiskit draws a flipped CNOT gate

Now, we'll look at SWAP gates.

SWAP gate

Another interesting two-qubit gate is a **SWAP gate**. As its name suggests, the SWAP gate exchanges the values of the two qubits. Here are two examples:

$$SWAP \ |01\rangle \ = |10\rangle$$

$$SWAP \left(\begin{pmatrix} \dfrac{1}{\sqrt{2}} \\ \dfrac{1}{\sqrt{2}} \end{pmatrix} \otimes \begin{pmatrix} \dfrac{\sqrt{3}}{2} \\ \dfrac{1}{2} \end{pmatrix} \right) = \begin{pmatrix} \dfrac{\sqrt{3}}{2} \\ \dfrac{1}{2} \end{pmatrix} \otimes \begin{pmatrix} \dfrac{1}{\sqrt{2}} \\ \dfrac{1}{\sqrt{2}} \end{pmatrix}$$

In each of these examples, you start with two qubits. Call them q_0 and q_1. After applying the SWAP gate, q_1 has the value that q_0 originally had, and q_0 has the value that q_1 originally had.

Figure 4.10 shows the effect of a SWAP gate on the values of two qubits:

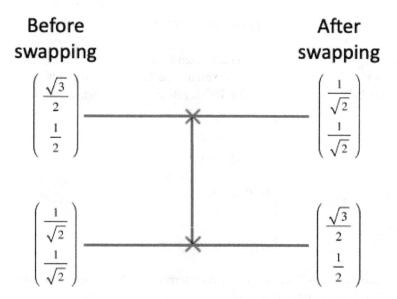

Figure 4.10 – Swapping two qubits' states

To swap qubits in Qiskit, call `circuit.swap(reg[0], reg[1])`. When Qiskit draws the SWAP gate, you see a vertical line with two Xs, as shown in *Figure 4.10*.

Let's look at one more multi-qubit gate.

Toffoli gate

Quantum computing doesn't stop with one- and two-qubit gates. A **controlled CNOT (CCNOT)** gate, also known as a **Toffoli gate**, operates on three qubits at a time:

```
from qiskit import QuantumRegister, QuantumCircuit
reg = QuantumRegister(3)
circuit = QuantumCircuit(reg)
circuit.ccx(reg[0], reg[1], reg[2])
display(circuit.draw('latex'))
```

Figure 4.11 shows the resulting circuit diagram:

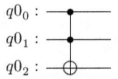

Figure 4.11 – A Toffoli (CCNOT) gate

The Toffoli gate in *Figure 4.11* applies NOT to the bottom qubit if and only if the top and middle qubits are both $|1\rangle$s. I don't know anyone who bothers to memorize the values in the Toffoli gate's matrix, but it's useful for you to have seen the matrix. The Toffoli gate matrix is shown as follows:

$$\begin{pmatrix} 1 & 0 & 0 & 0 & 0 & 0 & 0 & 0 \\ 0 & 1 & 0 & 0 & 0 & 0 & 0 & 0 \\ 0 & 0 & 1 & 0 & 0 & 0 & 0 & 0 \\ 0 & 0 & 0 & 0 & 0 & 0 & 0 & 1 \\ 0 & 0 & 0 & 0 & 1 & 0 & 0 & 0 \\ 0 & 0 & 0 & 0 & 0 & 1 & 0 & 0 \\ 0 & 0 & 0 & 0 & 0 & 0 & 1 & 0 \\ 0 & 0 & 0 & 1 & 0 & 0 & 0 & 0 \end{pmatrix}$$

When you apply this matrix to a vector, the matrix interchanges the values in the vector's $|111\rangle$ and $|011\rangle$ positions. That's how a Toffoli gate works.

In *Figure 4.11*, the bottom qubit is the gate's target qubit. You can switch around a Toffoli gate's registers so that the top or middle qubit is the target qubit. When you do this, you're changing the gate's matrix representation.

The next section reveals an advantage of multi-qubit gates.

Magic tricks with multi-qubit gates

Imagine this: You send two people to two different grocery stores. Before doing so, you give each an instruction to *buy either meat or fish*. You don't say which of the two products either person should buy. When they return from their respective stores, they both return with meat—not fish.

The next day, you do all this again in exactly the same way. "Buy either meat or fish," you say. On this second day, they both return with fish—not meat. On the third day, they both return with fish again. On the fourth day, they both return with meat. On a given day, you can't predict whether they'll both return with meat or both return with fish. But you know one thing for sure: they'll always return with the same food. What's more, if you repeat the experiment 100 times, they'll return approximately half the time with meat and the other half with fish.

Your first guess is that, before either person makes a purchase, the two people communicate in some way: "Hey, Jane. Let's both buy meat today." But if you rule out the possibility of any communication, how can you explain this spot-on coordination?

If you have two qubits, you can make them obey a similar kind of coordination. Of course, the qubits don't buy meat or fish. Instead, when you measure the qubits, either both of them are 0 or both of them are 1. This coordination between qubits helps you solve problems with quantum computers.

Introducing entanglement

Consider the circuit shown in *Figure 4.12*:

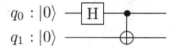

Figure 4.12 – The $CNOT(I \otimes H) |00\rangle$ circuit

When you run this circuit, something amazing happens. First, let's think informally about the way each of these gates works:

- A Hadamard gate turns $|0\rangle$ into the $|+\rangle$ state. The $|+\rangle$ is $\frac{1}{\sqrt{2}}(|0\rangle + |1\rangle)$. It's an equal superposition of $|0\rangle$ and $|1\rangle$. In a strange way, it's half $|0\rangle$ and half $|1\rangle$.

- A CNOT gate changes the target qubit's state from $|0\rangle$ to $|1\rangle$ if and only if the control qubit is $|1\rangle$.

With a control qubit halfway between $|0\rangle$ and $|1\rangle$, what happens to the target qubit? The answer is that the target qubit also goes halfway between $|0\rangle$ and $|1\rangle$. But the target qubit's halfway state isn't independent of the control qubit's halfway state. Here's how it works:

- If you measure any qubit, you always get either 0 or 1. You can't get anything in between 0 and 1.

- If you measure the two qubits after they've passed through the circuit in *Figure 4.12*, either both of them measure as 0 or both measure as 1.

  Informally, you can imagine that half of the control qubit is |0⟩ and the other half is |1⟩. The control qubit's "|1⟩ half" induces a change in the state of the target qubit, but the control qubit's "|0⟩ half" doesn't. (See *Figure 4.13*.)

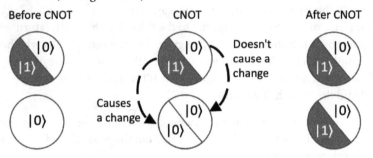

Figure 4.13 – An intuitive view of the CNOT gate's behavior

- But here's the strangest part of this circuit's behavior: before you measure either of the qubits, neither is *predestined* to be measured as either 0 or 1.

That last bullet is one of the most perplexing results in modern physics. In fact, no one knows exactly what's behind this phenomenon. After applying a Hadamard gate followed by a CNOT gate, you can have two qubits that dance in unison. The qubits are said to be **entangled**. They dance in unison, but they're not *coordinated* in our everyday sense of the word.

You start with two qubits in a laboratory in London. (See *Figure 4.14*.) The photo comes courtesy of Henry Be on Unsplash.

Figure 4.14 – Two qubits (both in the |0⟩ state)

You send one of the qubits through a Hadamard gate. (See *Figure 4.15*.)

Figure 4.15 – Two qubits (one $|+\rangle$ and the other $|0\rangle$)

You apply a CNOT gate with the $|+\rangle$ qubit as the control qubit. (See *Figure 4.16.*)

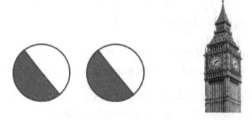

Figure 4.16 – Two entangled qubits

You send the two qubits to distant parts of the galaxy. (See *Figure 4.17.*)

Figure 4.17 – Two qubits (one in the Cygnus constellation, the other in Centaurus)

Some strange-looking creature in the Cygnus galaxy measures the Cygnus qubit, and another creature in the Centaurus galaxy measures the other qubit. If the Cygnusian gets 0 when measuring the qubit, the Centaurian also gets 0. If the Cygnusian gets 1, the Centaurian also gets 1.

So, what's going on here? The two qubits are 100,000 light-years apart, so they don't have time to tell each other about their measurement outcomes. Maybe, as the qubits parted ways from their position in London, they agreed on a common measurement outcome. "Let's both be 0 when we're measured," whispered one qubit to the other. In 1926, Max Born wrote a paper [1] suggesting that this *whisper* was a **hidden variable**—a property of entangled particles that cannot be directly observed. In 1935,

Einstein, Podolsky, and Rosen joined the discussion [2] by arguing that, without something such as hidden variables, quantum mechanics made no sense.

Well, if you can't observe a hidden variable, how can you possibly determine when it exists or not? That's the topic of this chapter's *Qubits don't plan ahead* section.

Entanglement with matrices

The previous section offers an intuitive view of entanglement. To justify that intuitive view, we calculate the result of applying the gates in *Figure 4.12*:

$$CNOT(I \otimes H)|00\rangle = \begin{pmatrix} 1 & 0 & 0 & 0 \\ 0 & 0 & 0 & 1 \\ 0 & 0 & 1 & 0 \\ 0 & 1 & 0 & 0 \end{pmatrix} \frac{1}{\sqrt{2}} \begin{pmatrix} 1 & 1 & 0 & 0 \\ 1 & -1 & 0 & 0 \\ 0 & 0 & 1 & 1 \\ 0 & 0 & 1 & -1 \end{pmatrix} \begin{pmatrix} 1 \\ 0 \\ 0 \\ 0 \end{pmatrix} = \begin{pmatrix} 1 & 0 & 0 & 0 \\ 0 & 0 & 0 & 1 \\ 0 & 0 & 1 & 0 \\ 0 & 1 & 0 & 0 \end{pmatrix} \frac{1}{\sqrt{2}} \begin{pmatrix} 1 \\ 1 \\ 0 \\ 0 \end{pmatrix} = \frac{1}{\sqrt{2}} \begin{pmatrix} 1 \\ 0 \\ 0 \\ 1 \end{pmatrix} \begin{matrix} |00\rangle \\ |01\rangle \\ |10\rangle \\ |11\rangle \end{matrix}$$

The end result is either $|00\rangle$ or $|11\rangle$. The $|01\rangle$ and $|10\rangle$ combinations have 0 amplitudes, so their probabilities are both 0. We give this result its own name. We call it $|\Phi^+\rangle$ (pronounced as *f-eye plus*).

The $|\Phi^+\rangle$ state contains two qubits, so it's natural to ask if we can describe each qubit's state independently of the other. Is the $|\Phi^+\rangle$ state a tensor product of two-qubit states?

The answer is a resounding *no*.

> **Remember this**
>
> Two qubits can be represented as a tensor product of vectors if and only if the qubits are not entangled.

In the language of physics, the states in an entangled pair are not **separable**.

Entangled or not, the sum of the probabilities must add up to 1. So, let's check that:

$$|\Phi^+\rangle = \frac{1}{\sqrt{2}} \begin{pmatrix} 1 \\ 0 \\ 0 \\ 1 \end{pmatrix} = \begin{pmatrix} \frac{1}{\sqrt{2}} \\ 0 \\ 0 \\ \frac{1}{\sqrt{2}} \end{pmatrix} \qquad \left| \frac{1}{\sqrt{2}} \right|^2 + \left| \frac{1}{\sqrt{2}} \right|^2 = \frac{1}{2} + \frac{1}{2} = 1$$

This calculation reminds us that, when we measure the two qubits, the chance of seeing one of the four possible outcomes ($|00\rangle$, $|01\rangle$, $|10\rangle$, or $|11\rangle$) is 100%. That's reassuring!

Working with Qiskit

In this section, we'll write code to create entangled qubits. We'll use a number of Qiskit features to visualize the qubits' entangled state. Follow these steps:

1. Create a new Qiskit notebook (follow the steps from *Chapter 1, New Ways to Think about Bits*).

2. In the first cell, type and run the following code:

```
from qiskit import QuantumCircuit

circ = QuantumCircuit(2)
circ.h(0)
circ.cnot(0, 1)
display(circ.draw('latex', initial_state=True))
```

The resulting circuit diagram is shown in *Figure 4.18*:

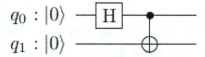

Figure 4.18 – A circuit that entangles qubits

3. In the next cell, type and run the following code:

```
from qiskit.quantum_info import Statevector

vector = Statevector(circ)
display(vector.draw(output='qsphere'))
```

The output we get is shown in *Figure 4.19*:

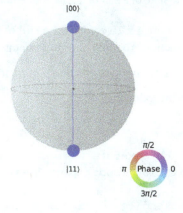

Figure 4.19 – Qiskit displays entangled qubits

In previous chapters, we called `plot_bloch_multivector(vector.data)` to display a qubit's Bloch sphere. You can do that in this section. But with entangled qubits, you can also call `vector.draw(output='qsphere')`. In *Figure 4.19*, each dot on the sphere represents a two-qubit state. One of these states is $|00\rangle$, and the other is $|11\rangle$. The line between the dots suggests entanglement. The circle below the sphere uses color to describe the phases of the two possibilities.

> **Important note**
>
> If you're reading a black-and-white copy of this book, you can't interpret the phase circle in *Figure 4.19*, so I'll guide you through it. The circle's rim has many different colors. Near label **0**, the rim is blue. In addition, the dots on the sphere are both blue. This indicates that the dots on the sphere both have phase 0.

In truth, the faithful representation of an entangled pair would require a Bloch sphere in four dimensions. I have enough trouble showing three-dimensional spheres on a two-dimensional page. I dare not ask you to visualize four dimensions. Anyway, the call to `draw(output='qsphere')` does a decent job of describing the two-qubit state.

If you don't like the color circle in *Figure 4.19*, the next step presents an alternative.

4.   In the next cell, type and run the following code:

```
display(vector.draw(output='latex'))
```

Figure 4.20 shows you the resulting output:

$$\frac{\sqrt{2}}{2}|00\rangle + \frac{\sqrt{2}}{2}|11\rangle$$

Figure 4.20 – The state of a pair of qubits after entanglement

This is $|\Phi^+\rangle$. It's equivalent to the 4 x 1 vector at the end of the matrix calculation in the *Entanglement with matrices* section. You just have to remember that $\frac{\sqrt{2}}{2}$ is the same as $\frac{1}{\sqrt{2}}$. (Some people would call my $\frac{1}{\sqrt{2}}$ an *improper fraction*. I don't see what's so improper about it.)

5.   In the next cell, type and run the following code:

```
circ.measure_all()

from qiskit import Aer, execute
from qiskit.visualization import import plot_histogram

device = Aer.get_backend('qasm_simulator')
job = execute(circ, backend=device, shots=1000)
```

```
counts = job.result().get_counts(circ)
print(counts)
plot_histogram(counts)
```

Figure 4.21 shows you the resulting output:

```
{'00': 516, '11': 484}
```

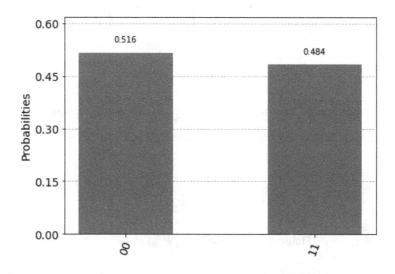

Figure 4.21 – Roughly half |00⟩ and half |11⟩

As predicted, a run of the circuit shows that both qubits are |0⟩ or both qubits are |1⟩. These qubits are married to one another.

The four Bell states

In the previous section, you entangled two qubits to create a combined $\frac{1}{\sqrt{2}}$ (|00⟩ + |11⟩) state. When you measure these two qubits, you're guaranteed to get the same value.

This combined $\frac{1}{\sqrt{2}}$ (|00⟩ + |11⟩) state is one of four possible **Bell states**. The circuit in *Figure 4.22* creates a different Bell state—one in which two qubits are guaranteed to have the *opposite* values when they're measured:

$$q_0 : |0\rangle \; -\boxed{H}-\bullet-$$
$$q_1 : |0\rangle \; -\oplus-\boxed{X}-$$

Figure 4.22 – Creating the Bell state $\frac{1}{\sqrt{2}}$ (|01⟩ + |10⟩)

The nickname for the $\frac{1}{\sqrt{2}}(|01\rangle + |10\rangle)$ state is $|\Psi^+\rangle$ (pronounced as *ps-eye plus*). The strategy in *Figure 4.22* is to entangle the two qubits and then change one of the qubits from $|0\rangle$ to $|1\rangle$ or from $|1\rangle$ to $|0\rangle$. It works!

The remaining two Bell states involve a phase difference between the two-qubit possibilities. *Figure 4.23* shows the circuit to create $\frac{1}{\sqrt{2}}(|00\rangle - |11\rangle)$, also known as $|\Phi^-\rangle$:

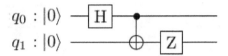

Figure 4.23 – Creating the state $|\Phi^-\rangle$

Rotating one of the qubits around the z axis throws the combination out of phase.

Figure 4.24 combines the strategies in *Figure 4.22* and *Figure 4.23* to create $\frac{1}{\sqrt{2}}(|01\rangle - |10\rangle)$, also known as $|\Psi^-\rangle$:

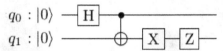

Figure 4.24 – Creating the $|\Psi^-\rangle$ state

So far in this chapter, you've learned what entanglement is and how to create entanglement using quantum circuits. Entanglement plays an important role in quantum computing, but with some algorithms, the benefit of entangling qubits isn't easy to understand. The next section presents a general overview of the use of entanglement in quantum algorithms.

Role of entanglement in quantum computing

For insight into the use of entanglement, let's take a quick glance at the circuit in *Figure 4.25*:

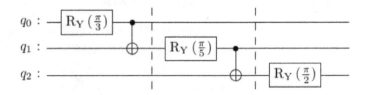

Figure 4.25 – Three-way entanglement

Without making any measurements, you know the following:

- The outcome of measuring q_0 will be either 0 or 1

- Before passing the first barrier, the eventual measurement of q_1 depends on the eventual measurement of q_0

- Before passing the second barrier, the eventual measurement of q_2 depends on the eventual measurement of q_1

- At the far right of the diagram, the eventual measurement of q_2 also depends on the application of that qubit's R_Y gate

Figure 4.26 shows a decision tree that keeps track of all these dependencies:

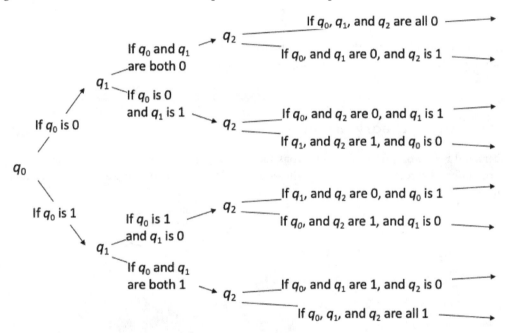

Figure 4.26 – Simulating three-way entanglement with classical branching

Looking at *Figure 4.26* too carefully will give you a headache! The takeaway from this diagram is that the number of branches grows exponentially with the number of qubits. In short, to simulate three qubits, a classical computer has to manage 2^3 different paths. That's much more than what's required to deal with a measly collection of three classical bits.

There's no question about it—entanglement is powerful. But how does entanglement work under the hood? How do two qubits coordinate their measurement outcomes? Is this something we can describe with our day-to-day understanding of things? The next section tackles that fascinating question.

Qubits don't plan ahead

In the 1999 movie *Mystery Men*, one character has the superpower of making himself invisible, but only when no one is looking. For most people, this raises the question, "What good is that superpower?" For me, it raises an entirely different question: "Since no one can witness this character disappearing, is there a way to find out if the character actually disappears?" Can you verify or disprove the existence of something that, by its very nature, is unobservable? Of course, the knee-jerk answer to this question is, "No, you can't."

But, in 1964, physicist John Bell wrote a paper [4] in which he proposed an experiment that could put an end to hidden-variable theories. Then, in 1982, the team of Alain Aspect, Philippe Grangier, and Gérard Roger performed a convincing version of Bell's experiment [5].

> **Disclaimer**
>
> This section introduces the theory about the nature of entanglement. If this theory doesn't interest you, feel free to skip it. None of this book's other material depends on the concepts in this *Qubits don't plan ahead* section.

What quantum theory predicts

To understand Bell's work, think about electrons approaching measurement devices that are oriented in several different directions. *Figure 4.27* shows an **up** electron approaching a similarly oriented measuring device:

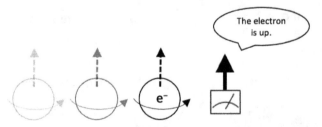

Figure 4.27 – An "up" electron meets an "up" device

When the electron in *Figure 4.27* reaches the measuring device, the device registers *up*.

Figure 4.28 shows a **down** electron approaching a downward-oriented measuring device:

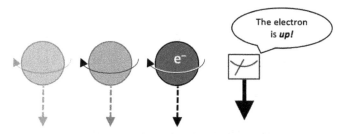

Figure 4.28 – A "down" electron meets a "down" device

Up is whatever you want it to be. Think about the meanings of *up* in England and Australia. *Up* means different things in different locations on Earth. When the electron in *Figure 4.28* reaches the measuring device, the device registers **up**.

Figure 4.29 shows an *up* electron approaching a measuring device that's oriented $\frac{\pi}{6}$ radians (30 degrees) from the electron's *up*:

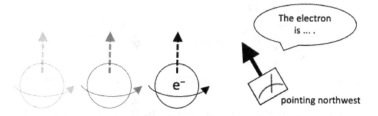

Figure 4.29 – An "up" electron approaches a device at an angle

The situation in *Figure 4.29* is identical to the one in *Figure 4.30*. As in all other scenarios, *up* is what you make of it:

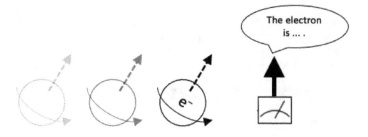

Figure 4.30 – An electron at an angle approaches an "up" device

In fact, the situation in *Figure 4.30* is what we have when we apply an $R_y\left(\frac{\pi}{6}\right)$ rotation. Using a formula from *Chapter 3, Math for Qubits and Quantum Gates*, (and looking up the values of the trig functions for $\frac{\pi}{12}$), we get the following:

$$R_Y\left(\frac{\pi}{3}\right)|0\rangle = \begin{pmatrix} \cos\dfrac{\pi}{12} & -\sin\dfrac{\pi}{12} \\ \sin\dfrac{\pi}{12} & \cos\dfrac{\pi}{12} \end{pmatrix}\begin{pmatrix} 1 \\ 0 \end{pmatrix} = \begin{pmatrix} \dfrac{\sqrt{3}+1}{2\sqrt{2}} & -\dfrac{\sqrt{3}-1}{2\sqrt{2}} \\ \dfrac{\sqrt{3}-1}{2\sqrt{2}} & \dfrac{\sqrt{3}+1}{2\sqrt{2}} \end{pmatrix}\begin{pmatrix} 1 \\ 0 \end{pmatrix} = \begin{pmatrix} \dfrac{\sqrt{3}+1}{2\sqrt{2}} \\ \dfrac{\sqrt{3}-1}{2\sqrt{2}} \end{pmatrix} \approx \begin{pmatrix} 0.966 \\ 0.259 \end{pmatrix}$$

The probability of measuring *up* is roughly $0.966^2 \approx 0.933$, and the probability of measuring *down* is approximately $0.259^2 \approx 0.067$.

In the same way, you can start with a device that's oriented $\dfrac{5\pi}{6}$ radians (150 degrees) from the electron's spin. (See *Figure 4.31*.)

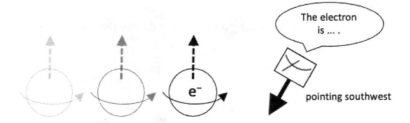

Figure 4.31 – An "up" electron approaches a device at yet another angle

The probabilities in *Figure 4.31* are the opposite of those in *Figure 4.30*. That is, the probability of measuring *up* is roughly 0.067, and the probability of measuring *down* is approximately 0.933.

Now, imagine that an electron meets a measuring device that's randomly set at one of three possible angles. (See *Figure 4.32*.)

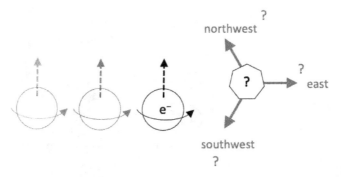

Figure 4.32 – An electron approaches a measuring device at one of three angles

For your reading pleasure, I've decided to name the measuring directions **northwest**, **southwest**, and **east** instead of $\frac{\pi}{6}$, $\frac{5\pi}{6}$, and $\frac{9\pi}{6}$. The probability of the measuring device being northeast-oriented is $\frac{1}{3}$. The same is true of the east and southwest orientations. We ask, what's the probability that the measuring device reports *up*? If your intuition answers $\frac{1}{2}$, your intuition is correct. Let's do the math. (What? You say you're not totally comfortable with math involving probabilities? Don't worry. There's a *Combining probabilities* section near the end of this chapter.)

First, we do an approximation, like so:

$P(up)$

$= P(\,northwest\,) \cdot P(\,up \mid northwest\,) \;+\; P(\,southwest\,) \cdot P(\,up \mid southwest\,) \;+\; P(\,east\,) \cdot P(\,up \mid east\,)$

$\approx \dfrac{1}{3} \cdot 0.993 \;+\; \dfrac{1}{3} \cdot 0.067 \;+\; \dfrac{1}{3} \cdot \dfrac{1}{2} \;\approx\; \dfrac{1}{2}$

If approximations don't convince you, here's the rigorous derivation:

$P(up)$

$= P(\,northwest\,) \cdot P(\,up \mid northwest\,) \;+\; P(\,southwest\,) \cdot P(\,up \mid southwest\,) \;+\; P(\,east\,) \cdot P(\,up \mid east\,)$

$= \dfrac{1}{3} \cdot \left(\dfrac{\sqrt{3}+1}{2\sqrt{2}} \right)^{2} \;+\; \dfrac{1}{3} \cdot \left(\dfrac{\sqrt{3}-1}{2\sqrt{2}} \right)^{2} \;+\; \dfrac{1}{3} \cdot \dfrac{1}{2}$

$= \dfrac{1}{3} \left(\dfrac{3 + 2\sqrt{3} + 1}{8} \;+\; \dfrac{3 - 2\sqrt{3} + 1}{8} \;+\; \dfrac{1}{2} \right)$

$= \dfrac{1}{2}$

To the rightward-moving electron in *Figure 4.32*, let's add a second, leftward-moving electron. (See *Figure 4.33*.) This leftward-moving electron has a downward spin and heads toward an east-oriented measuring device:

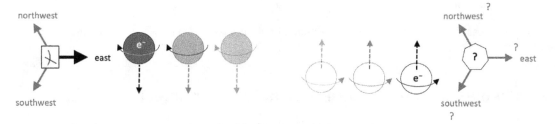

Figure 4.33 – Sending two electrons in opposite directions

Each measuring device reports either up or down. There's never anything in between. The question is, what's the probability that the two measuring devices disagree with one another? What's the probability that one of the devices says *up* and the other says *down*? Let's have a closer look:

- The analysis accompanying *Figure 4.32* tells us that the probability of getting *up* from the measuring device on the right is $\frac{1}{2}$. The probability of getting *down* is also $\frac{1}{2}$.

- On the left, the measuring device is oriented halfway between the approaching electron's *up* and *down* directions. So, the probability of getting *up* from the device on the left is $\frac{1}{2}$. The probability of getting *down* is also $\frac{1}{2}$.

Let's summarize:

$$P\,(disagreement)$$

$$= P\,(leftUp \text{ and } rightDown) \,+\, P\,(leftDown \text{ and } rightUp)$$

$$= P\,(leftUp) \cdot P\,(rightDown) \,+\, P\,(leftDown) \cdot P\,(rightUp)$$

$$= \frac{1}{2} \cdot \frac{1}{2} + \frac{1}{2} \cdot \frac{1}{2}$$

$$= \frac{1}{2}$$

In the scenario shown in *Figure 4.33*, the two measuring instruments disagree approximately half the time.

Figure 4.33 illustrates the situation in which the electron on the left has upward spin, and the measuring device on the left points eastward. The same kind of reasoning applies if we randomize the left side's electron spin and the measuring orientation.

In Bell's experiment, you randomize the measurement orientations on both the left and right sides. When you add up all the possibilities in Bell's experiment, the probability of having measurements that disagree is $\frac{1}{2}$.

> **Important note**
>
> For each case in Bell's experiment, the probability of disagreement is $\frac{1}{2}$. That's what quantum theory predicts.

What would happen if there were hidden variables?

At the end of the previous section, we concluded that the probability of having two measurements disagree in Bell's experiment is $\frac{1}{2}$. In this section, we come to a different conclusion based on the assumption that quantum mechanics has hidden variables. So, which of these contradicting conclusions is correct? You'll find out soon enough.

Let's imagine that quantum theory includes hidden variables—properties of entangled particles that cannot be directly observed. Two entangled qubits leave London on journeys that will take them 100,000 light-years apart. As the qubits part ways from their position in London, they agree on a common measurement outcome. "If we're eventually measured by northeasterly devices, I'll be *down*, and you'll be *up*," whispers one qubit to the other. As the qubits begin moving apart, one qubit has a variable with a value of *if northeast, then down*, and the other qubit has a variable with a value of *if northeast, then up*. In a sense, each qubit *knows* what value it will have if it's eventually measured by a northeasterly device. *Figure 4.34* illustrates this idea in more detail:

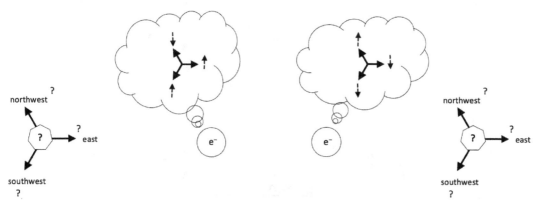

Figure 4.34 – Each qubit "knows" the outcome of its eventually being measured

In *Figure 4.34*, as each qubit races toward its measuring device in a distant part of the galaxy, the qubit *knows* what the outcome will be when it's eventually measured. For example, the qubit on the left knows that if it's measured by a northeasterly device, the measurement will be *down*. Because the two qubits are oppositely entangled, the qubit on the right has the opposite value. The qubit on the right knows that if it's measured by a northeasterly device, the measurement will be *up*.

Figure 4.35 shows possibilities for the other two measurement orientations, as follows:

- If both the left and right measurements are made in the easterly direction, the left qubit's measurement comes out to be *up*, and the right qubit's measurement comes out to be *down*

- If both the left and right measurements are made in the southwesterly direction, the left qubit's measurement comes out to be *up*, and the right qubit's measurement comes out to be *down*

In Bell's experiment, the left and right measuring devices don't have to be oriented the same way. For example, the left device may be northwesterly and the right device easterly. In fact, there are nine different possibilities. In *Figure 4.35*, we consider all nine of them:

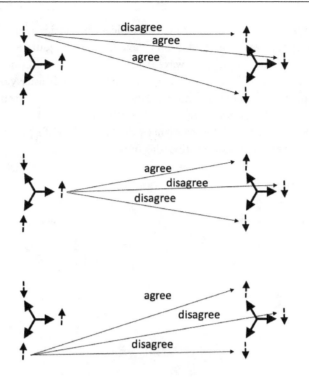

Figure 4.35 – Nine ways to select orientations for the left and right measurements

In *Figure 4.35*, the two measurements disagree for five of the nine possibilities. So, maybe, if there are hidden variables, the probability of disagreement isn't $\frac{1}{2}$. Instead, it's $\frac{5}{9}$.

But wait! The hidden variables don't have to point the way they do in *Figure 4.34*. Maybe they look like the rendition in *Figure 4.36*:

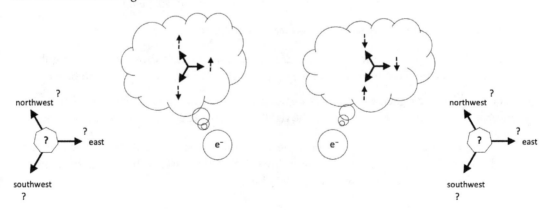

Figure 4.36 – Each qubit "knows" the outcome of its eventually being measured (version 2)

Figure 4.36 is almost the same as *Figure 4.34*. The only difference is that, in the thought bubbles, some of the arrows are reversed. Using those reversed arrows, you can work out all the agreements and disagreements, as I did in *Figure 4.35*. When you do, you get the same answer—namely, that the probability of disagreement is exactly $\frac{5}{9}$.

The only situation that doesn't yield $\frac{5}{9}$ is one in which, on either side, all three device orientations give you the same measurement. (See *Figure 4.37*.)

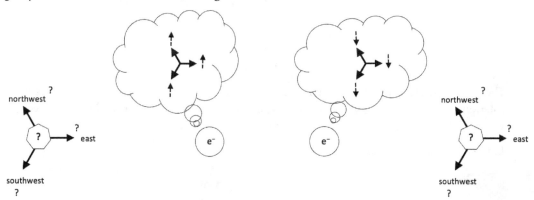

Figure 4.37 – No matter how the devices are oriented, the left and right measurements disagree

For a situation such as the one in *Figure 4.37*, the probability of disagreement is 1.

> **Important note**
>
> For each case in Bell's experiment, the probability of disagreement is at least $\frac{5}{9}$. That's a consequence of assuming that quantum theory has hidden variables.

But a probability of at least $\frac{5}{9}$ contradicts the value $\frac{1}{2}$ that we got in the *What quantum theory predicts* section. So, what's the real truth? Only experiments can say for sure. And in 1982, experiments performed by Aspect, Dalibard, and Roger (http://dx.doi.org/10.1103/PhysRevLett.49.1804) concluded that quantum theory is correct. There are no hidden variables in quantum mechanics.

Bell's experiment in Qiskit

In 1982, three scientists in France spent months setting up expensive equipment to perform Bell's experiment. But today, you can run the experiment from your own laptop computer. Here's how:

1. Start a new Qiskit notebook and run this code in the first cell:

```
from qiskit import QuantumCircuit
```

```
def get_circuit(angle_left, angle_right):
    circ = QuantumCircuit(2)
    circ.h(0)
    circ.cnot(0, 1)
    circ.x(1)
    circ.barrier()
    circ.ry(angle_left, 0)
    circ.ry(angle_right, 1)
    circ.measure_all()
    display(circ.draw('latex'))
    return circ
```

The get_circuit function returns a circuit in which two entangled qubits experience R_Y rotations. The angles of the rotations depend on the function's parameters. Since you haven't yet called the get_circuit function, a run of this cell's code creates no output.

2. In the next cell, run the following code:

```
from math import pi

northwest = pi / 6
southwest = 5 * pi / 6
east = 9 * pi / 6
directions = [northwest, southwest, east]

circuits = []
for dir_left in directions:
    for dir_right in directions:
        circuits.append(
            get_circuit(dir_left, dir_right))
```

This code calls the get_circuit function nine times—once for each combination of left and right measuring orientations. The code adds each circuit to a list of circuits. To no one's surprise, the list's name is circuits.

Upon each call to get_circuit, the notebook displays the newly created circuit. The resulting output is shown in *Figure 4.38*. In the diagram, notice how we simulate the measurements of qubits at various angles. Instead of rotating the two measurement gates, we rotate the two qubits. Either way, the end result is the same. It doesn't matter whether we rotate a qubit or its measuring device. What matters is the angle between these two things:

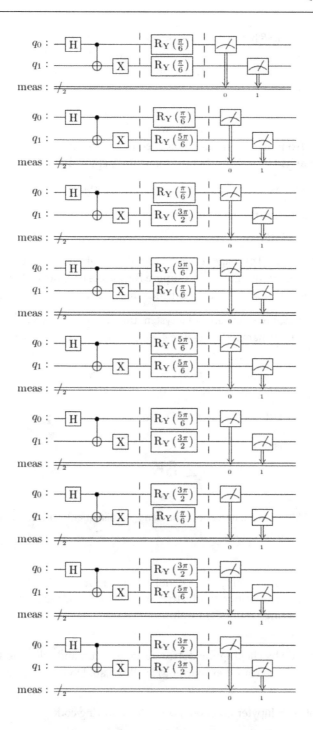

Figure 4.38 – The nine circuits in Bell's experiment

Each circuit in *Figure 4.38* has two qubits. So, in the next several steps, you find the name of a real quantum computer (not a simulator) with at least two available qubits. Believe it or not, some of the backends in IBM's quantum computing cloud have only one qubit (at least, this was true in mid-2022).

You could run Bell's experiment on a simulator, but experimenting with a simulator would be cheating! After all, a simulator behaves the way we humans believe that qubits behave. To get the story straight from nature's mouth, you need real qubits By the time you read this book, IBM may have changed its website, but the following steps worked in mid-2023.

3. Click the dotted-rectangle icon (⠿) in the upper-right corner of the **IBM Quantum** window.

    A panel appears along the right side of the screen.

    In that panel, select **Platform**.

    A new web page appears. This page serves as your dashboard on the IBM Quantum Platform. On this page, look for options marked **Compute Resources** or **Your systems**. When you select one of these options, you'll see a list of backends that you can use to run your job..

4. In the list of backends, note the name of a system that has at least two qubits and doesn't have the word simulator in its name. (See *Figure 4.39*.).

Figure 4.39 – Information about the ibmq_manila quantum computer

---

Tip

When you click on a quantum computer's panel, you see more details about that device. In particular, you can find the number of jobs that are currently waiting in the device's queue. Look for a device with a small number of waiting jobs.

---

5. Go back to your code by clicking the dotted-rectangle icon and selecting **Learning**. On the resulting page, click a tab (or other marking) labeled **Lab**. Doing so brings you back to your Jupyter notebook.

6. In the next cell of your Jupyter notebook, copy the following code:

```
from qiskit_ibm_provider import IBMProvider
```

```
from qiskit import execute

provider = IBMProvider()
device = provider.get_backend('ibmq_manila')
# For a real test, run on a quantum backend
#      with at least two qubits

shots=100
job = execute(circuits, backend=device,
              shots=shots, memory=True)
result = job.result()
```

7. In the code from *step 7*, replace ibmq_manila with the name of the system you chose in *step 5*.

8. Press *Shift + Enter* to run the new cell's code.

The code's execute call has some interesting features. For one thing, this is the book's first example in which the call's initial parameter is a list of circuits rather than a single circuit. The shots value is 100. So, when Qiskit runs the job, it performs 100 shots with the first circuit in the list, 100 shots with the second circuit in the list, and so on. In total, Qiskit does 900 shots. In this watered-down experiment, we don't really select the left and right measuring orientations at random. Instead, we systematically test 100 of each of the nine possible orientations.

> **Tip**
>
> In several ways, this section's code is simpler than a true test of Bell's hypothesis. To test orientations at random, we'd have to create one enormous job with IBMQJobManager and chop the job into smaller chunks, We'd randomly select orientations after each pair of qubits became entangled, not before they're entangled as in Figure 4.38. That would add complexity that has little to do with Bell's experiment.

Another feature of this code's execute call is the memory=True parameter. Without that parameter, the job's result would tell us the number of 0 and 1 outcomes for the first qubit and the number of 0 and 1 outcomes for the second qubit. From that information, we'd have no way of knowing the number of times the two qubits disagreed. The memory=True parameter tells Qiskit to preserve the outcome of each shot. So, in the next piece of code, we can count the shots in which the qubits disagree.

9. In the next cell, run the following code:

```
disagree = 0
for circ in circuits:
    memory = result.get_memory(circ)
    for meas in memory:
        if meas[0] != meas[1]:
            disagree += 1
```

```
print('\nProbability of disagreement: ', end='')
print(disagree / (9 * shots))
```

In this code, a call to `result.get_memory(circ)` retrieves the outcomes of all 100 shots for one of this experiment's circuits. If you want to see these outcomes, you can add a `print(memory)` call. With each such call, you get a list of 100 two-qubit measurements. Here's the list I got when I added `print(memory)` to one of my runs:

```
['11', '11', '01', '11', '01', '10', '01', '01', '01', '11',
'10', '01', '01', '10', '01', '11', '11', '11', '01', '01',
'01', '01', '11', '11', '01', '01', '01', '01', '11', '01',
'11', '01', '11', '01', '01', '01', '00', '01', '11', '00',
'11', '01', '01', '01', '01', '11', '01', '01', '11', '01',
'01', '11', '01', '01', '01', '01', '11', '11', '11', '11',
'11', '11', '00', '01', '01', '01', '01', '01', '01', '00',
'01', '01', '11', '01', '01', '01', '01', '01', '01', '11',
'11', '01', '01', '01', '01', '11', '01', '11', '11', '01',
'01', '01', '01', '11', '00', '01', '11', '11', '11', '11']
```

This cell's inner loop counts the number of measurements in which the qubits' values disagree.

This code's actual output is the proportion of shots that disagree after running all 900 circuits. It looks something like this:

```
Probability of disagreement: 0.52
```

The final probability value may not be 0.52, but whatever that number is, it's very unlikely to be larger than $\frac{5}{9}$. In fact, by repeating this kind of experiment many times with many different kinds of equipment, physicists have concluded that the value $\frac{1}{2}$, as we deduced in the *What quantum theory predicts* section, is correct. Quantum mechanics has no hidden variables.

Congratulations! You've analyzed a phenomenon that's impossible to observe and show that it doesn't exist.

Combining probabilities

In the section entitled *Qubits don't plan ahead*, we use rules of probability to form conclusions about entangled qubits. If you're not familiar with these rules, this section is for you.

An **outcome** is one possible result from a randomly-conducted experiment. For example, you shuffle a standard, 52-card deck of playing cards. Then, you close your eyes and select one of the cards. One outcome of this experiment is that you pick the nine of hearts. Another outcome is that you pick the queen of spades. All in all, this experiment has 52 outcomes.

An **event** is a set of outcomes in a randomly conducted experiment. Again, pick one card from a shuffled, 52-card deck. Picking a red card is an example of an event because picking a red card means picking a card from the set of all hearts and diamonds. Picking a face card (jack, queen, or king) is another event. Picking an even-numbered card is an event. Picking either a 2 of clubs, a 10 of diamonds, or a king of spaces is yet another event. If you must know, the pick-a-card experiment has 2^{52} (about $4\frac{1}{2}$ quadrillion) events.

Assuming that all outcomes are equally likely, we have the following formula:

$$\text{probability of obtaining a paricular outcome} = P(\text{outcome}) = \frac{\text{number of events in the outcome}}{\text{count of all events in the experiment}}$$

For example, in the pick-a-card experiment, the probability of picking a spade is $\frac{1}{4}$ because the number of spaces is 13, the number of cards is 52, and $\frac{13}{52} = \frac{1}{4}$. With the letter P meaning *probability*, we have the following:

$$P(\text{spade}) = \frac{1}{4}$$

With these definitions, the basic rules of probability go in lockstep with the four fundamental arithmetic operations: addition, subtraction, multiplication, and division.

Addition (either this event or that event)

If two events have no outcomes in common, the following applies:

$$P\left(\text{event } E_1 \text{ or event } E_2\right) = P\left(\text{event } E_1\right) + P\left(\text{event } E_2\right)$$

For example, in the pick-a-card experiment, with events E_1 = picking a face card and E_2 = picking an ace, the probability of picking either a face card or an ace is as follows:

$$P(\text{face card or ace}) = P(\text{face card}) + P(\text{ace}) = \frac{12}{52} + \frac{4}{52} = \frac{16}{52} = \frac{4}{13}$$

Subtraction (not this event)

The probability of an event not occurring is given by the following formula:

$$P(\text{not E}) = 1 - P(E)$$

For example, the probability of not picking a club is shown here:

$$P(\text{not club}) = 1 - P(\text{club}) = 1 - \frac{1}{4} = \frac{3}{4}$$

Multiplication (this event and that event)

When the occurrence of one event has no influence on the occurrence of another, we say that the two events are *independent*. If events E_1 and E_2 are independent, then the following applies:

$$P\left(E_1 \text{ and } E_2\right) = P\left(E_1\right) \cdot P\left(E_2\right)$$

For example, perform the pick-a-card experiment twice using two different shuffled decks. The outcome of the first pick has no influence on the outcome of the second pick. The probability of selecting the ace of spaces twice is shown here:

$$P\left(\text{ace of spaces from deck}_1 \text{ and ace of spaces from deck}_2\right) = \frac{1}{52} \cdot \frac{1}{52} = \frac{1}{2704}$$

Division (this event assuming that event)

We define *conditional probability* as the probability that one event will occur when we know for sure that another event has occurred. The symbol for conditional probability is the vertical line character (|). Here's the formula:

$$P\left(\text{event } E_1 \text{ given that } E_2 \text{ has occurred}\right) = P\left(E_1 \mid E_2\right) = \frac{P\left(E_1\right)}{P\left(E_2\right)}$$

For example, in the pick-a-card experiment, if you know for sure that you've picked a face card, you can find the probability that your face card is a king:

$$P(\text{king} \mid \text{face card}) = \frac{4/52}{12/52} = \frac{4}{12} = \frac{1}{3}$$

If you know the probabilities of two or more events, you can combine the events in various ways. To find the probability of a particular combination, you use this section's addition, subtraction, multiplication, and division rules.

Summary

When combined with other gates, a two-qubit CNOT gate can entangle qubits. An entangled pair behaves as a single unit in which the measurement of one qubit determines the outcome of measuring the other. Neither qubit exists independently in a state of its own. The two-qubit system cannot be represented as the tensor product of two single qubits.

The true nature of entanglement remains a mystery for physicists. Experiments indicate that the theory has no hidden variables, so the qubits don't *know* if they'll be 0s or 1s before they're measured. But at the time of each measurement, the qubits may be separated by many light-years. And yet, news of one qubit's measurement seems to travel instantaneously to inform the other qubit's measurement. No one knows exactly why this happens. As physicist Richard Feynman said, the best we can do is to *"Shut up and calculate"*.

In this chapter, we leveraged the fact that you can't copy a qubit. You can't start with one qubit and end up with two identical qubits. So, when you send a message using qubits, only one person can receive the message. If a malicious eavesdropper reads the message, the person who was supposed to receive the message gets nothing but gibberish. In this case, the wrong person reads the message, and the right person doesn't. What good is that? The next chapter answers that very question.

Questions

1. What's the output of the following circuit?

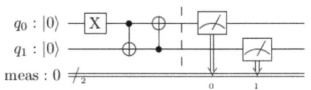

2. Try to find values of a_0, a_1, b_0, and b_1 satisfying $|\Phi^+\rangle = \begin{pmatrix} a_0 \\ a_1 \end{pmatrix} \otimes \begin{pmatrix} b_0 \\ b_1 \end{pmatrix}$. (It can't be done because the two qubits in question are entangled.)

3. Show that the circuit in *Figure 4.22* creates the $|\Phi^-\rangle$ state by writing the circuit's matrix representation and calculating the circuit's output.

4. Modify the code in this chapter's *Working with Qiskit* section so that the resulting state is $|\Psi^-\rangle$. Output a multi-qubit sphere like the one shown in *Figure 4.18*.

5. Between 1983 and 1993, there were approximately 105 boys born in the US for every 100 US-born girls. Given this ratio, what's the probability that a family of three would have exactly two girls and one boy (in no particular order)?

6. In this chapter's *What would happen if there were hidden variables?* section, we conclude that the overall probability of measurement disagreement is at least $\frac{5}{9}$. We say *at least* because this fraction accounts only for situations such as the one in *Figure 4.35*, not the one in *Figure 4.36*. Use the information accompanying *Figure 4.36* to make our conclusion more precise. That is, find a number, d, such that the probability of disagreement is *equal* to d.

Further readings

[1]. Born, M. (1926). Zur Quantenmechanik der Stossvorgänge. Zeitschrift für Physik, 37, 863-867. Translated as *On the quantum mechanics of collisions*, in *J. A. Wheeler* and *W. Zurek (eds), Quantum Theory and Measurement, Princeton, NJ: Princeton University Press (1983), pp. 52-55*

[2]. A. Einstein; B. Podolsky; N. Rosen (1935-05-15). Can Quantum-Mechanical Description of Physical Reality be Considered Complete? (PDF). *Physical Review. 47 (10): 777-780.* Bibcode:1935PhRv...47..777E. doi:10.1103/PhysRev.47.777

[3]. Bell, J. S. (1964). On the Einstein Podolsky Rosen Paradox (PDF). *Physics Physique Физика. 1 (3): 195-200.* doi:10.1103/PhysicsPhysiqueFizika.1.195

[4]. Aspect, A., Grangier, P., and Roger, G., 1982. Experimental realization of Einstein-Podolsky-Rosen-Bohm Gedankenexperiment: a new violation of Bell's inequalities, Physical Review Letters, 49: 91-94

Part 2
Making Qubits Work for You

In this part, we will use superposition and entanglement to perform useful tasks. One task is to exchange sensitive information in order to expose any eavesdroppers. Another task is to dismantle qubits and have their exact replicas appear elsewhere. The latter task is used in the low-level design of quantum computers and the implementation of network systems.

This part has the following chapters:

- *Chapter 5, A Fanciful Tale about Cryptography*
- *Chapter 6, Quantum Networking and Teleportation*

5

A Fanciful Tale about Cryptography

In our modern society, information privacy is essential. If people knew your credit card number, you could lose lots of money. If your medical records became public, a potential employer may decide not to hire you. If a foolish remark you made when you were ten years old circulated on social media sites, you could be banned from public office. And if a snooper catches you sending certain kinds of *sell all my shares* emails, you could go to jail for years and years.

Methods for sending secret messages have been around for millennia. One of the older methods is called the **Caesar Cipher** because it's named after the emperor himself – Julius Caesar. But secret messages can be intercepted and decrypted. One of the most famous cases took place during World War II when a team of researchers led by Alan Turing created a device to decrypt German military communications.

A common way to encrypt a message is to mix the message's content with some randomly generated content. This randomly generated content is called a **key**. Anyone who doesn't have that key value may have difficulty reversing the mixing process. So long as no malicious intruder discovers the key, your message is secure.

But this scheme of creating a key doesn't solve the security problem. After all, you need the key to make the message unreadable, and someone you trust needs that same key to read your message. Sending a key across a network may be as risky as sending an original message across a network. If you're worried about an intruder intercepting your original message, how do you know an intruder won't intercept the transmission of your key?

How can you distribute between a sender and a trusted receiver? Quantum computing has some clever ways of solving this problem. One of them is called **BB84**. It's an example of **Quantum Key Distribution (QKD)**. This chapter will show you how it works.

We'll cover the following topics in this chapter:

- How to share a key without sending it over a network

- Encrypting text with a secret key
- Why quantum key distribution works
- How to simulate the BB84 algorithm in Qiskit code

Technical requirements

The code for this chapter can be found online on GitHub: `https://github.com/ PacktPublishing/Quantum-Computing-Algorithms`.

Sharing secrets

Alice and Bob want to share a secret. They don't care what the secret means – the secret doesn't have to mean anything at all. The secret might be "Bob once spit in Alice's soup" or it might be "3*%Uw4YM^i44Sq." The content doesn't matter, so long as no one else knows it.

So, **Alice** randomly generates 20 bits and sends them to **Bob**. (See *Figure 5.1*.)

Bit number	Alice	Bob
1	1	1
2	0	0
3	1	1
4	0	0
5	0	0
6	1	1
7	0	0
8	0	0
9	1	1
10	1	1
11	1	1
12	1	1
13	0	0
14	1	1
15	1	1
16	1	1
17	0	0
18	1	1
19	0	0
20	0	0

Figure 5.1 – Alice sends bits to Bob

What they don't know is that an eavesdropper (named **Eve**) listens in on the line. **Eve** makes copies of the bits and sends them to **Bob**. (See *Figure 5.2*.)

Bit number	Alice	Eve	Bob
1	1	1	1
2	0	0	0
3	1	1	1
4	0	0	0
5	0	0	0
6	1	1	1
7	0	0	0
8	0	0	0
9	1	1	1
10	1	1	1
11	1	1	1
12	1	1	1
13	0	0	0
14	1	1	1
15	1	1	1
16	1	1	1
17	0	0	0
18	1	1	1
19	0	0	0
20	0	0	0

Figure 5.2 – Eve snoops

Sorry, Alice. Your bits aren't secret. Back to the drawing board!

Adding Hadamard gates

Alice knows all about quantum computing. So, before she sends her bits, she applies the **Hadamard** operator to each one. Since the Hadamard operator is its own inverse, **Bob** can recover each bit's original value by applying Hadamard to the qubit that he receives. (See *Figure 5.3*.)

Bit number	Alice			Bob	
1	1	H	$1/\sqrt{2}(|0\rangle - |1\rangle)$	H	1
2	0	H	$1/\sqrt{2}(|0\rangle + |1\rangle)$	H	0
3	1	H	$1/\sqrt{2}(|0\rangle - |1\rangle)$	H	1
4	0	H	$1/\sqrt{2}(|0\rangle + |1\rangle)$	H	0
5	0	H	$1/\sqrt{2}(|0\rangle + |1\rangle)$	H	0
6	1	H	$1/\sqrt{2}(|0\rangle - |1\rangle)$	H	1
7	0	H	$1/\sqrt{2}(|0\rangle + |1\rangle)$	H	0
8	0	H	$1/\sqrt{2}(|0\rangle + |1\rangle)$	H	0
9	1	H	$1/\sqrt{2}(|0\rangle - |1\rangle)$	H	1
10	1	H	$1/\sqrt{2}(|0\rangle - |1\rangle)$	H	1
11	1	H	$1/\sqrt{2}(|0\rangle - |1\rangle)$ ⟶	H	1
12	1	H	$1/\sqrt{2}(|0\rangle - |1\rangle)$	H	1
13	0	H	$1/\sqrt{2}(|0\rangle + |1\rangle)$	H	0
14	1	H	$1/\sqrt{2}(|0\rangle - |1\rangle)$	H	1
15	1	H	$1/\sqrt{2}(|0\rangle - |1\rangle)$	H	1
16	1	H	$1/\sqrt{2}(|0\rangle - |1\rangle)$	H	1
17	0	H	$1/\sqrt{2}(|0\rangle + |1\rangle)$	H	0
18	1	H	$1/\sqrt{2}(|0\rangle - |1\rangle)$	H	1
19	0	H	$1/\sqrt{2}(|0\rangle + |1\rangle)$	H	0
20	0	H	$1/\sqrt{2}(|0\rangle + |1\rangle)$	H	0

Figure 5.3 – Alice and Bob apply Hadamard operators

But, what's good for Bob, the goose, is good for Eve, the gander. By applying Hadamard before and after her measurement, **Eve** can read the original bits and send Alice's qubits faithfully onward to **Bob**. (See *Figure 5.4*.)

Bit number	Alice			Eve				Bob	
1	1	H	$1/\sqrt{2}(|0\rangle - |1\rangle)$	H	1	H	$1/\sqrt{2}(|0\rangle - |1\rangle)$	H	1
2	0	H	$1/\sqrt{2}(|0\rangle + |1\rangle)$	H	0	H	$1/\sqrt{2}(|0\rangle + |1\rangle)$	H	0
3	1	H	$1/\sqrt{2}(|0\rangle - |1\rangle)$	H	1	H	$1/\sqrt{2}(|0\rangle - |1\rangle)$	H	1
4	0	H	$1/\sqrt{2}(|0\rangle + |1\rangle)$	H	0	H	$1/\sqrt{2}(|0\rangle + |1\rangle)$	H	0
5	0	H	$1/\sqrt{2}(|0\rangle + |1\rangle)$	H	0	H	$1/\sqrt{2}(|0\rangle + |1\rangle)$	H	0
6	1	H	$1/\sqrt{2}(|0\rangle - |1\rangle)$	H	1	H	$1/\sqrt{2}(|0\rangle - |1\rangle)$	H	1
7	0	H	$1/\sqrt{2}(|0\rangle + |1\rangle)$	H	0	H	$1/\sqrt{2}(|0\rangle + |1\rangle)$	H	0
8	0	H	$1/\sqrt{2}(|0\rangle + |1\rangle)$	H	0	H	$1/\sqrt{2}(|0\rangle + |1\rangle)$	H	0
9	1	H	$1/\sqrt{2}(|0\rangle - |1\rangle)$	H	1	H	$1/\sqrt{2}(|0\rangle - |1\rangle)$	H	1
10	1	H	$1/\sqrt{2}(|0\rangle - |1\rangle)$	H	1	H	$1/\sqrt{2}(|0\rangle - |1\rangle)$	H	1
11	1	H	$1/\sqrt{2}(|0\rangle - |1\rangle)$	H	1	H	$1/\sqrt{2}(|0\rangle - |1\rangle)$ ⟶	H	1
12	1	H	$1/\sqrt{2}(|0\rangle - |1\rangle)$	H	1	H	$1/\sqrt{2}(|0\rangle - |1\rangle)$	H	1
13	0	H	$1/\sqrt{2}(|0\rangle + |1\rangle)$	H	0	H	$1/\sqrt{2}(|0\rangle + |1\rangle)$	H	0
14	1	H	$1/\sqrt{2}(|0\rangle - |1\rangle)$	H	1	H	$1/\sqrt{2}(|0\rangle - |1\rangle)$	H	1
15	1	H	$1/\sqrt{2}(|0\rangle - |1\rangle)$	H	1	H	$1/\sqrt{2}(|0\rangle - |1\rangle)$	H	1
16	1	H	$1/\sqrt{2}(|0\rangle - |1\rangle)$	H	1	H	$1/\sqrt{2}(|0\rangle - |1\rangle)$	H	1
17	0	H	$1/\sqrt{2}(|0\rangle + |1\rangle)$	H	0	H	$1/\sqrt{2}(|0\rangle + |1\rangle)$	H	0
18	1	H	$1/\sqrt{2}(|0\rangle - |1\rangle)$	H	1	H	$1/\sqrt{2}(|0\rangle - |1\rangle)$	H	1
19	0	H	$1/\sqrt{2}(|0\rangle + |1\rangle)$	H	0	H	$1/\sqrt{2}(|0\rangle + |1\rangle)$	H	0
20	0	H	$1/\sqrt{2}(|0\rangle + |1\rangle)$	H	0	H	$1/\sqrt{2}(|0\rangle + |1\rangle)$	H	0

Figure 5.4 – Eve snoops again

Introducing randomness

If Alice applies a Hadamard gate to each of her bits, Eve can read the bits and forward them to Bob. So, Hadamard gates alone don't make communications more secure.

But Alice has another idea. She randomly chooses whether or not to apply the Hadamard operation before sending each of her bits to **Bob**. (See *Figure 5.5*.)

Bit number	Alice			Eve	Bob	
1	1		$\lvert 1\rangle$	1	?	?
2	0	H	$1/\sqrt{2}(\lvert 0\rangle + \lvert 1\rangle)$	?	?	?
3	1	H	$1/\sqrt{2}(\lvert 0\rangle - \lvert 1\rangle)$	?	?	?
4	0	H	$1/\sqrt{2}(\lvert 0\rangle + \lvert 1\rangle)$	?	?	?
5	0		$\lvert 0\rangle$	0	?	?
6	1		$\lvert 1\rangle$	1	?	?
7	0	H	$1/\sqrt{2}(\lvert 0\rangle + \lvert 1\rangle)$	?	?	?
8	0		$\lvert 0\rangle$	0	?	?
9	1		$\lvert 1\rangle$	1	?	?
10	1	H	$1/\sqrt{2}(\lvert 0\rangle - \lvert 1\rangle)$	?	?	?
11	1	H	$1/\sqrt{2}(\lvert 0\rangle - \lvert 1\rangle)$	?	?	?
12	1		$\lvert 1\rangle$	1	?	?
13	0	H	$1/\sqrt{2}(\lvert 0\rangle + \lvert 1\rangle)$	?	?	?
14	1	H	$1/\sqrt{2}(\lvert 0\rangle - \lvert 1\rangle)$	?	?	?
15	1	H	$1/\sqrt{2}(\lvert 0\rangle - \lvert 1\rangle)$	?	?	?
16	1		$\lvert 1\rangle$	1	?	?
17	0	H	$1/\sqrt{2}(\lvert 0\rangle + \lvert 1\rangle)$	?	?	?
18	1		$\lvert 1\rangle$	1	?	?
19	0	H	$1/\sqrt{2}(\lvert 0\rangle + \lvert 1\rangle)$	?	?	?
20	0		$\lvert 0\rangle$	0	?	?

Figure 5.5 – Alice randomly decides when to apply the Hadamard operation

Without knowing which of the 20 bits have been *Hadamard'ed*, Eve has no way to measure the qubits and send copies to Bob. Doing so is physically and theoretically impossible. Unlike Dolly the Sheep, qubits cannot be cloned. This fact is the subject of the *You can't copy a qubit* section, later in this chapter.

But now we ask, "What has Alice gained?" Eve can't figure out whether Alice meant to send 0 or 1, but neither can Bob. The only way for Bob to decide what Alice meant to send would be for Alice to tell Bob which qubits she Hadamard'ed, and which she hadn't. But that would defeat the purpose. Eve could intercept that communication and discover the big secret. What can Alice do about that?

The answer is a clever scheme known as **BB84**, named after Charles Bennett and Gilles Brassard, who developed it in 1984.

Adding more randomness

Let's let both Alice and Bob make their own randomly generated decisions about applying or not applying the Hadamard operation. *Figure 5.6* shows what can happen when Eve doesn't interfere:

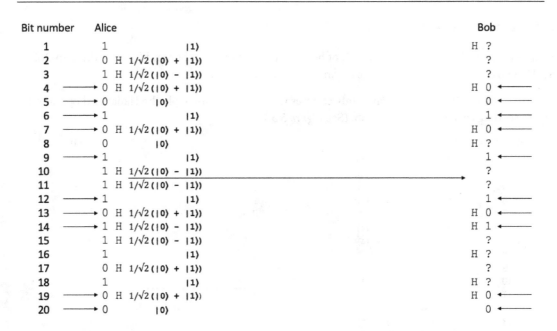

Figure 5.6 – Both Bob and Alice randomly decide when to apply Hadamard

With each of the 20 bits, there are two possibilities:

- Alice's and Bob's randomly generated decisions about applying Hadamard happen to agree. Either they both applied Hadamard, or neither applied Hadamard. (See the rows with short arrows in *Figure 5.6*.)

  In that case, Bob's measurement of Alice's bit is certain to be correct. Hadamard is its own inverse. So, two successive Hadamard operations give you a qubit's original, pristine state.

- Alice's and Bob's randomly generated decisions about applying Hadamard don't agree. One applied the Hadamard operation, and the other didn't.

  In that case, all bets are off. With only one Hadamard operation between Alice and Bob, the qubit is in the $\frac{1}{\sqrt{2}}(|0\rangle + |1\rangle)$ or $\frac{1}{\sqrt{2}}(|0\rangle - |1\rangle)$ state. So, no matter what Alice intends to transmit, Bob's measurement is either 0 or 1 with equal probability.

Alice and Bob perform the actions shown in *Figure 5.6*. Then, Bob takes out an ad in the New York Times announcing the choices he made before measuring each qubit. (See *Figure 5.7*.)

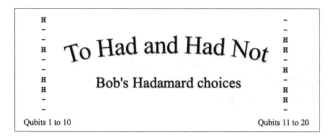

Figure 5.7 – Bob reveals his random sequence

Bob's flagrant disregard for any secrecy is not harmful at all. If Eve had been listening in, the information in Bob's ad would be of no use to her.

Alice and Bob compare some of their bits

As a next step in the BB84 algorithm, Alice rents time on one of Times Square's digital billboards. In this display, she announces which of her Hadamard choices agree with Bob's. (See *Figure 5.8*.)

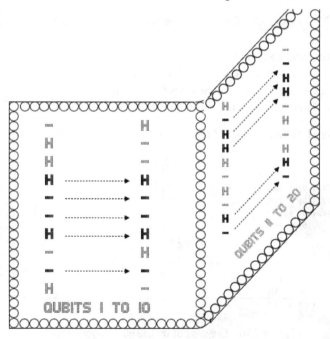

Figure 5.8 – 10 Hadamard agreement qubits

In *Figure 5.8*, Alice tells everyone that she and Bob agree on 10 of the qubits – namely, qubits 4, 5, 6, 7, 9, 12, 13, 14, 19, and 20. Let's call these the **Hadamard agreement qubits**.

Again, this new information would be of no use to an eavesdropper. There's no need to hide it from the rest of the world.

As a final gesture, while sharing information, Bob divides the Hadamard agreement qubits into two halves. He gives the first half (qubits 4, 5, 6, 7, and 9) a new name. He calls them the **test qubits**. Bob hires a skywriter to display the values he got when he measured each of these test qubits. (See *Figure 5.9*.)

Figure 5.9 – Bob's measurements of the test qubits

Alice sees the skywriting and compares Bob's measurements with the bits she intended to send. (See Alice's bits on the left-hand side of *Figure 5.6*.) For each of these five bits, Bob correctly measured the value of Alice's original bit. So, Alice concludes that no one was eavesdropping on their original 20-qubit transmission. How can Alice be so sure of that?

To answer the question, let's rework half of *Figure 5.6* with **Eve** in the middle. (See *Figure 5.10*.)

Bit number	Alice			Eve		Bob	Bit number
1	1		$\lvert 1 \rangle$	1		H ?	1
2	0	H	$1/\sqrt{2}(\lvert 0 \rangle + \lvert 1 \rangle)$	?		?	2
3	1	H	$1/\sqrt{2}(\lvert 0 \rangle - \lvert 1 \rangle)$	?		?	3
4	→ 0	H	$1/\sqrt{2}(\lvert 0 \rangle + \lvert 1 \rangle)$	?		H ? ←	4
5	→ 0		$\lvert 0 \rangle$	0		0 ←	5
6	→ 1		$\lvert 1 \rangle$	1		1 ←	6
7	→ 0	H	$1/\sqrt{2}(\lvert 0 \rangle + \lvert 1 \rangle)$	?		H ? ←	7
8	0		$\lvert 0 \rangle$	0		H ?	8
9	→ 1		$\lvert 1 \rangle$	1		1 ←	9
10	1	H	$1/\sqrt{2}(\lvert 0 \rangle - \lvert 1 \rangle)$	?		?	10
11	1	H	$1/\sqrt{2}(\lvert 0 \rangle - \lvert 1 \rangle)$	?		?	11

Figure 5.10 – Eve snoops on the BB84 algorithm

Notice Bob's question mark in the bit **4** row. This row is a Hadamard agreement row. So, Bob's measurement should accurately reflect the 0 that Alice intended to send in this row. But in this row, Bob's column contains a question mark. Alice applied a Hadamard gate. So, when Eve measured the qubit, she may have gotten 0 or 1. Because of this, we can't be sure that Bob's measurement will yield 0.

The same reasoning holds for the bit **7** row in *Figure 5.10*. If **Eve** is in the picture, there's little chance that **Bob** can accurately measure all five of the test qubits. So after *Figure 5.9*, when Alice looked at the skywriting and saw that Bob's bit values matched her bit values, she could be sure that Eve hadn't been listening in.

Some remaining bits form the secret key

With these Hadamard choices and test qubit values exposed to the public, where's the secret that Alice and Bob have been trying to generate? Bob announced the successful measurement of five bits with skywriting in *Figure 5.9*, so those bits aren't secret. But remember: Alice originally sent 20 bits to Bob, and bits **11** to **20** remain more or less untouched. Both Alice and Bob revisit the billboard in *Figure 5.8* and conclude that they have the same values for bits **12**, **13**, **14**, **19**, and **20**. That settles it! Their mutually shared secret is 10100 – the value of those five bits.

And the good news is, Alice never sent that five-bit sequence to Bob. Alice never said, "Let's make 10100 be our secret," so no one could have listened in and discovered this precious, life-changing secret.

Is the BB84 algorithm useful?

In the previous section, Alice and Bob formed a randomly generated secret. Both Alice and Bob know 10100 or whatever other sequence their BB84 algorithm generated. But what good is that? If Alice can't decide exactly what she wants to say to Bob, why should she bother saying anything at all?

The answer lies in what Alice does with the randomly generated secret. By combining her randomly generated secret with some meaningful information, Alice can form a sequence of characters that's meaningful to Bob but meaningless to an eavesdropper.

Here's some terminology:

- The randomly generated secret is called a **key**.

  For example, in the previous section, Alice and Bob created the key 10100.

- The meaningful information that must be kept from prying eyes is called **plaintext**.

  Imagine that Alice wants to send the word Stop! to Bob. Alice doesn't want anyone else to read this message. The word Stop! is an example of plaintext.

  There are many ways to represent a word such as Stop!. One way is to use the **Extended American Standard Code for Information Interchange** (**EASCII**). This code assigns an 8-bit string to each of the 256 characters. The letter S is 01010011, the letter t is 01110100, and so on. The EASCII encoding of Stop! is as follows:

  ```
  01010011 01110100 01101111 01110000 00100001
  ```

  In the world of computing, 01010011 01110100 01101111 01110000 00100001 is plaintext. Anyone with access to this sequence of bits can interpret it as Stop!.

- How you combine the key with the plaintext is called a **cipher**.

Good ciphers are complicated. But in this example, we will illustrate the idea with a simple application of the **exclusive or (XOR)** operation. *Figure 5.11* describes the XOR operation:

$$\begin{array}{cccc}
0 & 0 & 1 & 1 \\
\text{XOR } \underline{0} & \text{XOR } \underline{1} & \text{XOR } \underline{0} & \text{XOR } \underline{1} \\
0 & 1 & 1 & 0
\end{array}$$

Figure 5.11 – Applying XOR to two bits

The exclusive or operator's motto is, "My output is 1, so long as my first input is 1 *or* my second input is 1, but they're *not both* 1."

Table 5.1 shows what happens when you apply XOR to corresponding bits in the EASCII encoding of Stop! and eight copies of the 10110 key:

	Binary	Text
Plaintext	01010011 01110100 01101111 01110000 00100001	Stop!
Repeated key	10100**101** 00**101001** **01001010** **01010010** 10**10100**	
After applying XOR	11110110 01011101 00100101 00100010 10110101	ö]%"µ

Table 5.1 – Using XOR to scramble a message

- The unreadable sequence of characters you have after combining the key with the plaintext is called **ciphertext**.

When Alice applies XOR between Stop! and the repeated key, she creates the string of bits in the last row of *Table 5.1*. Anyone who interprets this string as EASCII text finds that the meaning of the string is ö] % "µ. That's nothing but a jumble of characters. If Eve is listening in, she can't make sense of it.

- The whole process of creating ciphertext from plaintext is called **encryption**.

When Alice encrypts the word Stop!, she gets ö] % "µ.

Alice's effort is an example of **secret key encryption**. It's *secret* because, for it to work, the 10100 key must be kept under wraps. Alice and Bob use BB84 to ensure that Eve doesn't know the key.

- The process of recovering the plaintext from the ciphertext is called **decryption**.

Recovering the plaintext is Bob's job. With our little XOR cipher, Bob doesn't have to work hard. When he applies XOR to the ciphertext and the key, he gets the original plaintext. (See *Table 5.2*.)

	Binary	Text
Ciphertext	11110110 01011101 00100101 00100010 10110101	ö]%"µ
Repeated key	10100**101** 00101001 **01001010** **01010010** 10**010100**	
After applying XOR	01010011 01110100 01101111 01110000 00100001	Stop!

Table 5.2 – Using XOR to unscramble a message

Another name for this kind of encryption is **symmetric encryption**. It's *symmetric* because the key for encrypting text is the same as the key for decrypting text.

And what about Eve? She doesn't have the key because Alice didn't send the key to Bob. Instead, Alice and Bob agreed on a key by sharing information on their Hadamard choices. So, Alice's precious `Stop!` message is safe and secure.

Or is it? Just for fun, I wrote a program for Eve to run. The program guesses keys and applies each resulting repeated key to the ciphertext. With my program, Eve can decrypt the message after only three thousandths of a second. What's wrong with this picture?

For encryption to work, we have to make two adjustments to the scheme described in this section. First, our proposed key with only 5 bits is much too short. In a typical symmetric encryption algorithm, the key must be at least 128 bits long.

Next, the XOR technique for combining the plaintext and the key is too simple. Real-life encryption algorithms involve XOR, plaintext bit rearranging, plaintext bit substitution, and other techniques.

With a sufficiently large key and a mathematically sound way of combining the key with the plaintext, you can establish very secure communication between a sender and a receiver.

You can't copy a qubit

The BB84 algorithm works because no eavesdropper can make a copy of a qubit's state. Imagine that Eve intercepts one of Alice's qubits, makes a measurement, and gets a value of 1. Eve has no way of knowing whether the qubit she measured was in the $|1\rangle$ state, the $\frac{1}{\sqrt{2}}(|0\rangle + |1\rangle)$ state, the $\frac{1}{\sqrt{2}}(|0\rangle - |1\rangle)$ state, or some other exotic in-between state. So, Eve doesn't know exactly what to forward to Bob.

But wait! Can we be sure that Eve has no way to make a copy of Alice's qubit? Yes, we can. The **No-Cloning theorem** shows that assuming that qubits can be copied leads to a nasty contradiction.

Let's start by agreeing on three properties of tensor products:

- For any three matrices, x, y, and z, a left distributive law holds – that is,
  $x \otimes (y + z) = (x \otimes y) + (x \otimes z)$.

You can write out a formal proof of this fact, but I always like to test with a simple example:

$$\begin{pmatrix} 1 \\ 2 \end{pmatrix} \otimes \left(\begin{pmatrix} 3 & 4 \\ 5 & 6 \end{pmatrix} + \begin{pmatrix} 7 & 8 \\ 9 & 0 \end{pmatrix} \right) = \begin{pmatrix} 1 \\ 2 \end{pmatrix} \otimes \begin{pmatrix} 10 & 12 \\ 14 & 6 \end{pmatrix} = \begin{pmatrix} 10 & 12 \\ 14 & 6 \\ 20 & 24 \\ 28 & 12 \end{pmatrix}$$

$$\left(\begin{pmatrix} 1 \\ 2 \end{pmatrix} \otimes \begin{pmatrix} 3 & 4 \\ 5 & 6 \end{pmatrix} \right) + \left(\begin{pmatrix} 1 \\ 2 \end{pmatrix} \otimes \begin{pmatrix} 7 & 8 \\ 9 & 0 \end{pmatrix} \right) = \begin{pmatrix} 3 & 4 \\ 5 & 6 \\ 6 & 8 \\ 10 & 12 \end{pmatrix} + \begin{pmatrix} 7 & 8 \\ 9 & 0 \\ 14 & 16 \\ 18 & 0 \end{pmatrix} = \begin{pmatrix} 10 & 12 \\ 14 & 6 \\ 20 & 24 \\ 28 & 12 \end{pmatrix}$$

An example is never as good as proof, but an example helps us build our intuitions.

- For any three matrices, x, y, and z, a right distributive law holds – that is, $(x + y) \otimes z = (x \otimes z) + (y \otimes z)$.

Make up an example to build your intuitions.

- For any scalar value, c, and any two matrices, x and y, the following law holds:

$$c\,(x \otimes y) = (cx) \otimes y = x \otimes (cy)$$

Here is an example:

$$5 \left(\begin{pmatrix} 1 \\ 2 \end{pmatrix} \otimes \begin{pmatrix} 3 & 4 \\ 5 & 6 \end{pmatrix} \right) = 5 \begin{pmatrix} 3 & 4 \\ 5 & 6 \\ 6 & 8 \\ 10 & 12 \end{pmatrix} = \begin{pmatrix} 15 & 20 \\ 25 & 30 \\ 30 & 40 \\ 50 & 60 \end{pmatrix}$$

$$\left(5 \begin{pmatrix} 1 \\ 2 \end{pmatrix} \right) \otimes \begin{pmatrix} 3 & 4 \\ 5 & 6 \end{pmatrix} = \begin{pmatrix} 5 \\ 10 \end{pmatrix} \otimes \begin{pmatrix} 3 & 4 \\ 5 & 6 \end{pmatrix} = \begin{pmatrix} 15 & 20 \\ 25 & 30 \\ 30 & 40 \\ 50 & 60 \end{pmatrix}$$

$$\begin{pmatrix} 1 \\ 2 \end{pmatrix} \otimes \left(5 \begin{pmatrix} 3 & 4 \\ 5 & 6 \end{pmatrix} \right) = \begin{pmatrix} 1 \\ 2 \end{pmatrix} \otimes \begin{pmatrix} 15 & 20 \\ 25 & 20 \end{pmatrix} = \begin{pmatrix} 15 & 20 \\ 25 & 30 \\ 30 & 40 \\ 50 & 60 \end{pmatrix}$$

As mentioned in *Chapter 4, Qubit Conspiracy Theories*, we can combine a circuit's qubits with a tensor product. For example, when we write $|1\rangle|0\rangle$ or $|10\rangle$, we're using shorthand to express $|1\rangle \otimes |0\rangle$. So, let's express the three tensor product rules in this shorthand notation. For three states, $|x\rangle$, $|y\rangle$, and $|z\rangle$, and any scalar value, c:

$$|x\rangle(|y\rangle + |z\rangle) = |x\rangle|y\rangle + |x\rangle|z\rangle = |xy\rangle + |xz\rangle$$

$$(|x\rangle + |y\rangle)|z\rangle = |x\rangle|z\rangle + |y\rangle|z\rangle = |xz\rangle + |yz\rangle$$

$$c|xy\rangle = c(|x\rangle|y\rangle) = |cx\rangle|y\rangle = |x\rangle|cy\rangle$$

To see these rules in action, think about a circuit with two qubits – one in the $\frac{1}{\sqrt{2}}|0\rangle + \frac{1}{\sqrt{2}}|1\rangle$ state and the other in the $\frac{1}{2}|0\rangle + \frac{\sqrt{3}}{2}|1\rangle$ state. We can describe the circuit by writing the following:

$$\left(\frac{1}{\sqrt{2}}|0\rangle + \frac{1}{\sqrt{2}}|1\rangle\right)\left(\frac{1}{2}|0\rangle + \frac{\sqrt{3}}{2}|1\rangle\right)$$

Then, we can multiply the terms using our tensor product distributive laws:

$$\left(\frac{1}{\sqrt{2}}|0\rangle + \frac{1}{\sqrt{2}}|1\rangle\right)\left(\frac{1}{2}|0\rangle + \frac{\sqrt{3}}{2}|1\rangle\right)$$

$$= \frac{1}{\sqrt{2}}|0\rangle\frac{1}{2}|0\rangle + \frac{1}{\sqrt{2}}|0\rangle\frac{\sqrt{3}}{2}|1\rangle + \frac{1}{\sqrt{2}}|1\rangle\frac{1}{2}|0\rangle + \frac{1}{\sqrt{2}}|1\rangle\frac{\sqrt{3}}{2}|1\rangle$$

$$= \frac{1}{2\sqrt{2}}|00\rangle + \frac{\sqrt{3}}{2\sqrt{2}}|01\rangle + \frac{1}{2\sqrt{2}}|10\rangle + \frac{\sqrt{3}}{2\sqrt{2}}|11\rangle$$

The bottom line describes a two-qubit circuit because the squares of the amplitudes add up to 1:

$$\left|\frac{1}{2\sqrt{2}}\right|^2 + \left|\frac{\sqrt{3}}{2\sqrt{2}}\right|^2 + \left|\frac{1}{2\sqrt{2}}\right|^2 + \left|\frac{\sqrt{3}}{2\sqrt{2}}\right|^2 = \frac{1}{8} + \frac{3}{8} + \frac{1}{8} + \frac{3}{8} = 1$$

In *Figure 5.12*, we are cloning a qubit:

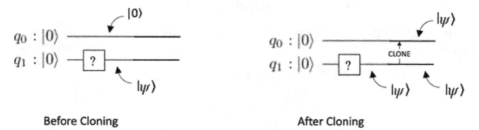

Before Cloning   After Cloning

Figure 5.12 – Copying a qubit's state

We start with two qubits – both in the $|0\rangle$ state. The bottom qubit passes through a gate that changes the qubit's state into some other state, $|\psi\rangle$. We don't particularly care what the $|\psi\rangle$ state is, but let's assume that $|\psi\rangle$ isn't $|0\rangle$ or $|1\rangle$. Instead, $|\psi\rangle$ is a superposition of some kind. After the bottom qubit passes through the gate, we have $|\psi\rangle|0\rangle$. Later in that same circuit, the bottom qubit encounters a cloning gate. After cloning, both qubits are in the $|\psi\rangle$ state.

We started by assuming that we can build a CLONE gate. Here's the formula that describes the gate's behavior:

$$Clone(|\psi\rangle|0\rangle) = |\psi\rangle|\psi\rangle$$

Like all qubit states, the $|\psi\rangle$ state is some combination of the $|0\rangle$ and $|1\rangle$ states:

$$|\psi\rangle = \alpha|0\rangle + \beta|1\rangle$$

After substituting this into the *Clone* formula, we can use our tensor product rules to cross-multiply:

$$
\begin{aligned}
Clone(|\psi\rangle|0\rangle) &= |\psi\rangle|\psi\rangle \\
&= (\alpha|0\rangle + \beta|1\rangle)(\alpha|0\rangle + \beta|1\rangle) \\
&= \alpha|0\rangle\alpha|0\rangle + \alpha|0\rangle\beta|1\rangle + \beta|1\rangle\alpha|0\rangle + \beta|1\rangle\beta|1\rangle \\
&= \alpha^2|0\rangle|0\rangle + \alpha\beta|0\rangle|1\rangle + \alpha\beta|1\rangle|0\rangle + \beta^2|1\rangle|1\rangle \\
&= \alpha^2|00\rangle + \alpha\beta|01\rangle + \alpha\beta|10\rangle + \beta^2|11\rangle
\end{aligned}
$$

On the last line, you may be tempted to combine $\alpha\beta|01\rangle$ with $\alpha\beta|10\rangle$. If so, remember that $|01\rangle$ isn't the same as $|10\rangle$. Just think about circuit diagrams. In $|01\rangle$, the bottom qubit is in the $|0\rangle$ state. But in $|10\rangle$, the bottom qubit is in the $|1\rangle$ state.

One way or another, the last line in the preceding equation tells us that, when we measure these two qubits, we may get $|01\rangle$ or $|10\rangle$. We won't always get $|00\rangle$ or $|11\rangle$. The only way that the last line would rule out $|01\rangle$ and $|10\rangle$ would be if either $\alpha = 0$ or $\beta = 0$. But then $|\psi\rangle$, which equals $\alpha|0\rangle + \beta|1\rangle$, would be either $|0\rangle$ or $|1\rangle$. We'd be trying to clone something that's essentially a classical bit. Let's not worry about that case. After all, a classical bit can be cloned. While I compose this sentence, my laptop is busy making copies of thousands upon thousands of classical bits.

Now, let's do the cloning equations again. This time, we'll expand $|\psi\rangle|0\rangle$ before we apply the CLONE operator:

$$
\begin{aligned}
Clone(|\psi\rangle|0\rangle) &= Clone\big((\alpha|0\rangle + \beta|1\rangle)|0\rangle\big) \\
&= Clone\big(\alpha|0\rangle|0\rangle + \beta|1\rangle|0\rangle\big) \\
&= \alpha Clone(|0\rangle|0\rangle) + \beta Clone(|1\rangle|0\rangle) \\
&= \alpha|0\rangle|0\rangle + \beta|1\rangle|1\rangle \\
&= \alpha|00\rangle + \beta|11\rangle
\end{aligned}
$$

When you clone $|0\rangle$, you get $|0\rangle|0\rangle$, while when you clone $|1\rangle$, you get $|1\rangle|1\rangle$. However, notice that contrary to the result we got in our earlier calculation, this new calculation offers no possibility of getting $|01\rangle$ or $|10\rangle$.

The two ways of expanding $Clone(|\psi\rangle|0\rangle)$ contradict one another, so performing CLONE must be impossible. That's the No-Cloning theorem. If Eve tries to faithfully copy the qubit coming from Alice's computer, Eve will fail. So, if Eve comes between Alice and Bob, her presence can be detected. That's the essence of the BB84 algorithm.

Qiskit code for the BB84 algorithm

Our BB84 simulation is a 150-line program (give or take a few lines). To make the program comprehensible, we'll divide it into 16 function definitions. Let's start with the imports and constant declarations:

```
import random
from qiskit import QuantumCircuit, QuantumRegister, \
    ClassicalRegister, Aer, execute


NUMBER_OF_CIRCUITS = 100
DOES_EVE_EXIST = False
CHECK_MARK = u'\u2713'
```

According to these declarations, we'll be creating 100 circuits – one for each of the qubits that Alice sends to Bob. We'll simulate a situation in which no eavesdropper exists. For convenience, we will declare CHECK_MARK to be the Unicode $\checkmark$ symbol.

The main flow of execution looks like this:

```
circuits = create_circuits(NUMBER_OF_CIRCUITS,
                           DOES_EVE_EXIST)           # 1

result = run_the_job(circuits)                       # 2

print_alice_bits(circuits)                           # 3

print_bob_bits(circuits, result)                     # 4

number_of_agreements = print_had_agreements(circuits) # 5

how_many_for_testing, is_eve_detected = \
    print_bit_agreements(circuits, result,
                         number_of_agreements)        # 6

if is_eve_detected:                                  # 7
    print('INTRUDER ALERT!')
else:
    print_key (circuits, NUMBER_OF_CIRCUITS,
               how_many_for_testing)
```

I've numbered the code's statements for easy reference. Refer back to the code's statement numbers as you read the explanations that follow:

- Statement 1 creates 100 circuits. In each circuit, the software randomly decides whether Alice generates 0 or 1, whether or not Alice applies the Hadamard gate before sending the value onward, and whether or not Bob applies the Hadamard gate before measuring the qubit. Since `DOES_EVE_EXIST` is `False`, none of the circuits contains any eavesdropping gates.

- Statement 2 runs all hundred circuits on a simulator or a real quantum computer.

- Statement 3 displays Alice's randomly generated sequence of zeros and ones. Each zero or one belongs to one of the program's 100 circuits. *Figure 5.13* shows what you might see in your notebook:

```
alice bits: 00011001111111010101101110000011101110000000011100001011100110011010111100000100100100101011000011
bob bits  : 00011001110111100101101110000011100110000100011100001010000011011010110001000110010000011111110001011
hads agree? ✓  //////  ///  / //   //  ///// / /  /  ////   ////   // / // /// /    // /// // // / ////  /
bits agree? ✓  //////  ///  / //   //  ///// / /  /  ////
key       :                                            0010   00 1 01 010 1    00 100 00  01 1  1000   1
```

Figure 5.13 – Alice's randomly generated bits

- Statement 4 displays the values that Bob gets when Bob measures the bits. If either Alice or Bob applies a Hadamard gate and the other person doesn't, the value that Bob measures may not be the same as Alice's value:

```
alice bits: 00011001111111010101101110000011101110000000011100001011100110011010111100000100100100101011000011
bob bits  : 00011001110111100101101110000011100110000100011100001010000011011010110001000110010000011111110001011
hads agree? ✓  //////  ///  / //   //  ///// / /  /  ////   ////   // / // /// /    // /// // // / ////  /
bits agree? ✓  //////  ///  / //   //  ///// / /  /  ////
key       :                                            0010   00 1 01 010 1    00 100 00  01 1  1000   1
```

Figure 5.14 – Bob gets these values when he measures the qubits

- Statement 5 displays a checkmark below each Hadamard agreement circuit. The display in your notebook will look like the last line in *Figure 5.15*:

```
alice bits: 00011001111111010101101110000011101110000000011100001011100110011010111100000100100100101011000011
bob bits  : 00011001110111100101101110000011100110000100011100001010000011011010110001000110010000011111110001011
hads agree? ✓  //////  ///  / //   //  ///// / /  /  ////   ////   // / // /// /    // /// // // / ////  /
bits agree? ✓  //////  ///  / //   //  ///// / /  /  ////
key       :                                            0010   00 1 01 010 1    00 100 00  01 1  1000   1
```

Figure 5.15 – Checkmarks indicate circuits in which the choices about applying H gates agree

- Statement 6 designates half of the Hadamard agreement circuits to be the test circuits. This statement puts an additional checkmark below each test circuit in which Alice's and Bob's bits agree. In *Figure 5.16*, 27 circuits get these additional checkmarks:

```
alice bits: 000110011111110101011011100000111011100000000111000010111001100110101111100000100100100101011000011
bob bits  : 000110011101111001011011100000111001100001000111000010100000110110101100010001100100000111111110001011
hads agree? ✓ ////// /// / // // ////// / / / //// //// // / // /// / // /// // // / //// /
bits agree? ✓ ////// /// / // // ////// / / / ////
key       :                                                      0010   00 1 01 010 1    00 100 00  01 1  1000   1
```

Figure 5.16 – Checkmarks indicate bit agreement between Alice and Bob

Statement 6 also sets the value of `is_eve_detected`. The value is `True` if all 27 circuits get the additional checkmarks and `False` otherwise.

- If no eavesdropper was detected, statement 7 displays the values of the remaining Hadamard agreement bits. *Figure 5.17* displays 28 circuits' bit values. These bits form the secret encryption key:

```
alice bits: 000110011111110101011011100000111011100000000111000010111001100110101111100000100100100101011000011
bob bits  : 000110011101111001011011100000111001100001000111000010100000110110101100010001100100000111111110001011
hads agree? ✓ ////// /// / // // ////// / / / //// //// // / // /// / // /// // // / //// /
bits agree? ✓ ////// /// / // // ////// / / / ////
key       :                                                      0010   00 1 01 010 1    00 100 00  01 1  1000   1
```

Figure 5.17 – Alice and Bob can safely use the 001000101010010000001110001 encryption key

In real life, the code wouldn't display an encryption key. Instead, Alice's and Bob's computers would simply use the same key bits based on their discovery that Eve, the eavesdropper, doesn't exist.

If we had set the `DOES_EVE_EXIST` constant to `True` in the notebook's first cell, the run wouldn't create an encryption key. *Figure 5.18* shows a sample run of this kind:

```
alice bits: 111001000101010011100010110100011100001111010110010000010101101010011111100101111000010100011101110
bob bits  : 111001010101010000010101011010000110000110010101101111000101010001101111101001001010010101011011001110
hads agree? //////// /     // / / //  /// / // /    /   /  / // //// //     // / / /     / /  ///////// /
bits agree? /////// /     // X
```

INTRUDER ALERT!

Figure 5.18 – Alice's and Bob's test bits don't all agree

The preceding paragraphs describe the code's main flow of execution. In that main flow, we describe what each function call does without showing any function's code. In the remainder of this chapter, we'll cover each function's code.

Creating the circuits

The `create_circuits` function returns a list of circuits. The number of circuits depends on the function's `how_many` parameter, and the kind of circuit depends on the value of the Boolean `does_eve_exist` parameter:

```
def create_circuits(how_many, does_eve_exist):
    circuits = []
    for i in range(how_many):
```

```
        circuits.append(make_new_circuit(does_eve_exist))
    return circuits
```

The `make_new_circuit` function creates a single circuit:

```
def make_new_circuit(eve_exists):
    circ = create_registers(eve_exists)
    alice_q = circ.qubits[0]
    bob_q = circ.qubits[1]
    bob_c = circ.clbits[0]

    circ = setup_alice(circ)

    circ.swap(alice_q, bob_q)

    if eve_exists:
        circ = setup_eve(circ)

    circ = setup_bob(circ)

    return circ
```

The `make_new_circuit` function calls `create_registers` to design the basic outline of the circuit:

```
def create_registers(eve_exists):
    alice_q = QuantumRegister(1, 'alice_q')
    bob_q = QuantumRegister(1, 'bob_q')
    bob_c = ClassicalRegister(1, 'bob_c')

    if eve_exists:
        eve_c = ClassicalRegister(1, 'eve_c')
        circ = QuantumCircuit(alice_q, bob_q, bob_c, eve_c)
    else:
        circ = QuantumCircuit(alice_q, bob_q, bob_c)
    return circ
```

The call to `create_registers` gives us one of the circuits shown in *Figure 5.19*:

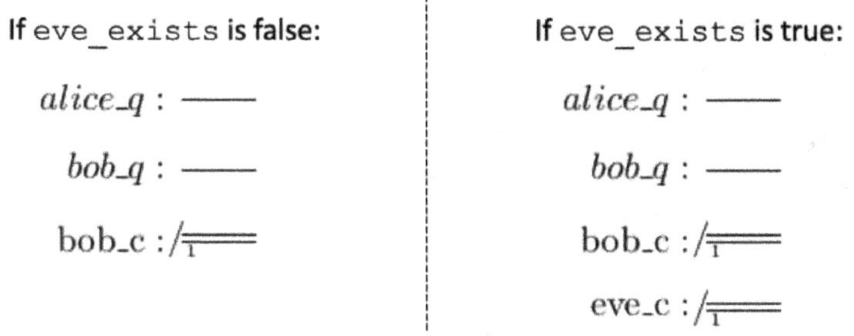

If `eve_exists` **is false:**

$$alice\_q : \text{———}$$

$$bob\_q : \text{———}$$

$$bob\_c : /\!\!=\!\!=$$

If `eve_exists` **is true:**

$$alice\_q : \text{———}$$

$$bob\_q : \text{———}$$

$$bob\_c : /\!\!=\!\!=$$

$$eve\_c : /\!\!=\!\!=$$

Figure 5.19 – Two kinds of circuits before gates are added

After calling `create_registers`, the `make_new_circuit` function does the following:

- It adds Alice's gates by calling `setup_alice`
- It adds Bob's gates by calling `setup_bob`
- Depending on a randomly generated value, it may add Eve's gates by calling `setup_eve`

Two of these three `setup_` functions have their own randomly chosen options:

```
def setup_alice(circ):
    alice_q = circ.qubits[0]

    if random.getrandbits(1):
        circ.x(alice_q)

    if random.getrandbits(1):
        circ.h(alice_q)

    return circ


def setup_bob(circ):
    bob_q = circ.qubits[1]
    bob_c = circ.clbits[0]

    if random.getrandbits(1):
        circ.h(bob_q)

    circ.measure(bob_q, bob_c)
    return circ


def setup_eve(circ):
```

```
bob_q = circ.qubits[1]
eve_c = circ.clbits[1]

circ.barrier()
circ.measure(bob_q, eve_c)
circ.barrier()
return circ
```

The result of each `make_new_circuit` call is a circuit with at least two and at most six gates. The *fat* arrows in *Figure 5.20* point to a SWAP gate and Bob's measurement gate – the only two gates that always appear when we call `make_new_circuit`:

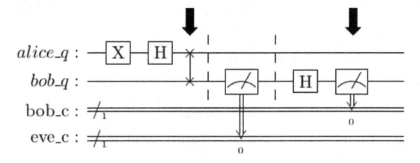

Figure 5.20 – The largest circuit that a make_new_circuit call can return

In addition to Alice's, Bob's, and possibly Eve's gates, the `make_new_circuit` function creates a SWAP gate. This SWAP gate is completely unnecessary. It represents my effort to simulate the passing of a qubit from Alice's hardware to Bob's hardware. If you want, you can eliminate the SWAP gate by combining Alice's and Bob's wires into one wire.

This completes our discussion of the code to create circuits. The next few sections will describe other aspects of the BB84 code.

Running the quantum circuits

In the functions that we've defined so far, we created 100 quantum circuits – one for each bit that Alice sends to Bob. Here's the code that runs all the circuits:

```
def run_the_job(circuits):
    device = Aer.get_backend('qasm_simulator')
    job = execute(circuits, backend=device, shots=1, memory=True)
    return job.result()
```

The code to run the circuits isn't very exciting. It runs each circuit once and retains each run's memory.

Displaying the outcome

Figure 5.17 shows the output of a typical run of this chapter's code. In the following sections, we'll describe the functions that generate each row in the output.

Alice's bits

The top row in *Figure 5.17* displays the zeros and ones that Alice randomly generates. How do we determine which circuits contain zeros and which contain ones?

Qiskit's `QuantumCircuit` class has a `count_ops` method. Here's one possible output of a call to `circ.count_ops()`:

```
OrderedDict([('h', 2), ('swap', 1), ('measure', 1)])
```

A call to `count_ops` returns an ordered dictionary consisting of gate names and gate counts. The output shown here describes a circuit with two Hadamard gates and no X gates. From this, we can draw two conclusions:

- Alice and Bob both applied Hadamard gates, so this is a Hadamard agreement circuit
- Alice didn't apply an X gate, so the bit she sent was 0

From the output of `count_ops`, we can also decide whether Eve's additional measurement gate is present in the circuit. But, in practice, that information isn't explicitly available to Alice and Bob. So, we will ignore that information in this code.

To display the zeros and ones that Alice randomly generates, we can check each circuit's `count_ops` dictionary for the presence or absence of an X gate:

```
def print_alice_bits(circuits):
    print('alice bits: ', end='')
    for circ in circuits:
        bit = 1 if 'x' in circ.count_ops() else 0
        print(bit, end='')
    print('')
```

Bob's bits

The second row in *Figure 5.17* displays each of Bob's measurements. As we did in *Chapter 4, Qubit Conspiracy Theories*, we get these measurements from the memory of each circuit's result:

```
def print_bob_bits(circuits, result):
    print('bob bits   : ', end='')
    for circ in circuits:
        memory = result.get_memory(circ)
        print(bob_bit_value(circ, memory), end='')
    print('')
```

In this code, the `bob_bit_value` function does the nitty-gritty work of extracting the relevant bit from the circuit's memory:

```
def bob_bit_value(circ, memory):
    return memory[0][0]
```

Hadamard agreement

The third line in *Figure 5.17* has checkmarks for each of the Hadamard agreement circuits. To find these circuits, we can look for `count_ops` dictionaries whose number of Hadamard gates is 0 or 2. Here's the code that checks the Hadamard agreement for a particular circuit:

```
def had_agreement(circ):
    gate_counts = circ.count_ops()
    return not ('h' in gate_counts and gate_counts['h'] == 1)
```

Here's the code that loops through all the circuits to create the entire Hadamard agreement row:

```
def print_had_agreements(circuits):
    number_of_agreements = 0
    print('hads agree? ', end='')
    for circ in circuits:
        if had_agreement(circ):
            print(CHECK_MARK, end='')
            number_of_agreements += 1
        else:
            print(' ', end='')
    print('')
    return number_of_agreements
```

Bit agreement

The fourth line of output of *Figure 5.17* sets aside half of the Hadamard agreement circuits as test circuits. For each test circuit, the line has a checkmark wherever Bob's measurement is the same as the value that Alice intended to send:

```
def print_bit_agreements(circuits, result,
                         number_of_agreements):
    number_tested = 0
    is_eve_detected = False
    i = 0

    print('bits agree? ', end='')
    while number_tested < number_of_agreements // 2:
        if had_agreement(circuits[i]):
```

```
            if bit_value_agreement(circuits[i], result):
                print(CHECK_MARK, end='')
                number_tested += 1
            else:
                is_eve_detected = True
                print('X')
                break
        else:
            print(' ', end='')
        i += 1

    print()

    return i, is_eve_detected
```

The call to `bit_value_agreement` compares the return values from two other functions:

```
def bit_value_agreement(circ, result):
    memory = result.get_memory(circ)
    return alice_bit_value(circ) == int(
        bob_bit_value(circ, memory))
```

We get Alice's bit value by counting the number of X gates in the circuit:

```
def alice_bit_value(circ):
    return 1 if 'x' in circ.count_ops() else 0
```

Finding an encryption key

The final lines of output in *Figure 5.17* and *Figure 5.18* show either an encryption key's bits or a message, warning that security is being compromised. The logic for constructing this output is in a function named `print_key`:

```
def print_key(circuits, number_of_circuits, how_many_for_testing):
    print('key        :', end='')
    for i in range(how_many_for_testing + 1):
        print(' ', end='')
    for i in range(i, NUMBER_OF_CIRCUITS):
        if had_agreement(circuits[i]):
            print(alice_bit_value(circuits[i]), end='')
        else:
            print(' ', end='')
```

This ends our tour of the code to simulate the BB84 algorithm. You can read the code in detail, or you can look at the code's higher-level functions and trust the other functions to do what their names suggest. One way or another, you may learn about some features of Qiskit that are useful in other contexts. You can download the code by visiting `https://github.com/PacktPublishing/Quantum-Computing-Algorithms`.

Getting more information about a circuit

In the previous section, we used the `QuantumCircuit` class's `count_ops` function to find out whether Alice applies an X gate and whether the circuit has an even or odd number of Hadamard gates. Sometimes, you need to discover more details about an existing circuit. For cases of this kind, you can use the `QuantumCircuit` class's `data` attribute. A circuit's `data` attribute contains enough information to recreate the circuit in its entirety. Take, for example, the circuit shown in *Figure 5.21*:

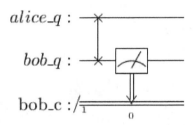

Figure 5.21 – The smallest circuit returned by any call to make_new_circuit

If you print this circuit's `data` attribute, and you add your own line breaks, you will see the following information:

```
[
    CircuitInstruction(
      operation=Instruction(
        name='swap', num_qubits=2, num_clbits=0, params=[]
      ),
      qubits=(
        Qubit(QuantumRegister(1, 'alice_q'), 0),
        Qubit(QuantumRegister(1, 'bob_q'), 0)
      ),
      clbits=()
    ),
    CircuitInstruction(
      operation=Instruction(
        name='measure', num_qubits=1, num_clbits=1, params=[]
      ),
      qubits=(Qubit(QuantumRegister(1, 'bob_q'), 0),),
      clbits=(Clbit(ClassicalRegister(1, 'bob_c'), 0),)
    )
]
```

This sequence describes two `Instruction` instances (swap and measure), two `QuantumRegister` instances (`alice_q` and `bob_q`), and one `ClassicalRegister` instance (`bob_c`). From this sequence, you can learn anything you want to know without looking at the diagram in *Figure 5.21*.

Summary

Using classical communications, any information that you send over a network can be intercepted and read by a malicious agent. This includes meaningful information such as your credit card number, but it also includes any randomly generated key that you use to encrypt the meaningful information. One way to achieve secure message transmission is for both the sender and the receiver to *have* information without ever transmitting that information over a network. We know of no way to do this with credit card numbers or any other meaningful information. But, using the BB84 algorithm, the sender and receiver can cooperatively create a random key that's known only to the two of them. This random key is never transmitted along network lines.

The BB84 algorithm depends on one important fact: you can't clone a qubit. If you get a qubit in some arbitrary state, $\alpha|0\rangle + \beta|1\rangle$, you can't measure the values of α and β to end up with two qubits in the same $\alpha|0\rangle + \beta|1\rangle$ state. But what if you're willing to destroy the first qubit? Can you make one qubit relinquish its state to another qubit? And, if you can, how close must the two qubits be to one another? Can the qubits be light-years apart? The next chapter examines these questions and will give you some surprising new answers.

Questions

1. *Figure 5.6* contains lots of little details. Keep me honest by checking to make sure that the H letters, question marks, zeros, ones, and arrows are all placed correctly on Bob's side of the figure.

2. Repeat your work from *Question 1*, this time verifying the ones, zeros, and question marks in Eve's column in *Figure 5.10*.

3. In carrying out the BB84 algorithm, Alice and Bob share some information out in the open. Bob shares all of his Hadamard choices, Alice announces which qubits are Hadamard agreement qubits, and Bob displays half of his Hadamard agreement bits. Justify the claim that making this information public doesn't help Eve discover the key.

4. Make up an example to build your intuitions about the right distributive law for tensor products: $(x + y) \otimes z = (x \otimes z) + (y \otimes z)$.

5. We began this chapter's *You can't copy a qubit* section by stating three laws for tensor products. Do we use all three of these laws in our argument about the truth of the No-Cloning theorem?

6. Write Qiskit code to create a circuit of the following kind:

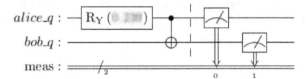

I've blurred out the angle of rotation for Alice's qubit because you should generate this value randomly.

Run this circuit several times to convince yourself that Bob's qubit ends up in the same state as Alice's qubit. To further convince yourself, calculate the state of Bob's qubit using matrix arithmetic.

With Bob's qubit in the same state as Alice's qubit, why don't we claim that this circuit clones Alice's qubit?

7. In this chapter, we assumed that Eve never applies any Hadamard gates. That's okay because having Eve apply Hadamard gates doesn't change the overall outcome.

The BB84 algorithm works whether Eve applies Hadamard gates or not. As an exercise, modify this chapter's code so that, for each circuit, Eve randomly decides whether she will or will not apply Hadamard gates. (If she decides to apply Hadamard gates, she applies two of them – one before measuring the qubit she receives from Alice, and another before sending the qubit on to Bob.)

8. Eve tries to break the BB84 algorithm by entangling Alice's qubit with another qubit and sending that other qubit onward to Bob. This fiendish plan doesn't work. (If it did, this whole chapter would be worthless!) Why doesn't Eve's plan work?

6

Quantum Networking and Teleportation

In 2022, a team of scientists in China "teleported" qubits from a station on Earth to an Earth-orbiting satellite several hundred miles away. After putting each qubit in a superposition state, they used entanglement to "beam" the qubits' states to a laboratory on the satellite. In the end, the original Earthly qubits had been measured and forced into state $|0\rangle$ or $|1\rangle$, but the satellite's qubits were in the same superposition states that the team had prepared on Earth. In effect, the state of affairs on Earth had been moved to the lab on the satellite.

A century from now, we may be teleporting human beings. Alice disappears here on Earth and suddenly appears alongside Bob on an orbiting satellite. "Where am I?" asks Alice. "How did I get here?" And with a wry smile, Bob says, "It's nice to meet you in person after our long-distance relationship in *Chapter 5!*"

This chapter covers the strange phenomenon of quantum teleportation, including these topics:

- Transmitting bits and qubits
- Teleporting a qubit
- Teleportation versus cloning
- Coding the teleportation circuitry

Technical requirements

The code used in this chapter can be found online on GitHub: `https://github.com/PacktPublishing/Quantum-Computing-Algorithms`.

Transmitting bits and qubits

Imagine a computer with no networking capabilities. You can compose a document, but you can't send it anywhere. You can't visit websites. You can't send or receive messages. If your computer runs a commercial application, you can't access new data without plugging in a thumb drive. No doubt about it. Networking is important.

Quantum networking is the transmission of qubit states from one device to another. This transmission opens up new possibilities. Here are some examples:

- The BB84 algorithm covered in *Chapter 5* involves the exchange of a randomly generated sequence of qubits. This requires a network that supports the transmission of qubit states.

- Today's cloud computing is endlessly scalable, and fast, reliable network communication makes cloud computing possible. When computers pool resources, the whole is more than the sum of its parts. A network of quantum computers would offer mind-boggling possibilities.

- Sensors in separate locations can fine-tune their readings by sharing quantum information. If telescopes positioned around the Earth could share quantum information, the combined system could form a super telescope – one large enough to view details on the surfaces of distant bodies. Alien beings, wave for the camera!

What kinds of challenges do we face before we can communicate quantum information? How does quantum networking compare with today's ubiquitous classical networking technology? The following sections have some answers to these questions.

Classical networks

Bits are sturdy. We've moved bits from one place to another for several decades. You can argue about the history, but a significant breakthrough came in 1969 when a group at the US government's **Advanced Research Projects Agency Network** (**ARPANET**) defined the technology that governs today's internet. Fast forward to the year 2022, when the worldwide median speed for mobile data downloads was about 30 million bits per second [1]. That's a lot of bits.

When you send bits from place to place, these bits travel over a **medium** of some kind. In the 1980s, much of the world transmitted bits over ordinary telephone wires. Today, if you're using Wi-Fi, the medium is radio waves, and many companies provide internet access through the medium of optical fiber.

No matter what medium you use, the signals that represent classical bits degrade over long distances. But don't worry. This degradation can be managed. Much of today's information flows over **Ethernet** cables, which adhere to a standard named **IEEE 802.3**. The standard defines the characteristics of a fast and reliable signal. Cables vary in length, but all else being equal, a short cable is better than a longer cable. With a cable that's longer than 100 meters, IEEE 802.3 makes no promises. "That cable is too long," says a trusty networking expert. That's why classical computer networking uses repeaters.

A **repeater** is a device that sits between a signal's source and the signal's destination. For an Ethernet cable, a repeater amplifies a degraded signal and restores the signal to IEEE 802.3 specifications. These days, repeaters are everywhere. You can buy one online for about 45 USD. (Of course, you may need to borrow your neighbor's repeater in order to get online and make the purchase!)

Figure 6.1 presents a simplified view of a classical repeater's action.

> **Important note**
> In real-life networking, the lines in parts **(a)** and **(c)** are never as clean as they appear in *Figure 6.1*. But for the purpose of this discussion, that doesn't matter.

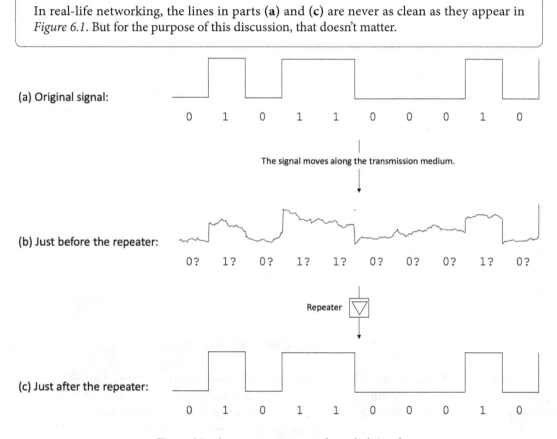

Figure 6.1 – A repeater restores a degraded signal

If you look only at the degraded signal in part **(b)** of *Figure 6.1*, you can guess where the zeros and ones are. A repeater makes guesses of this kind and turns its guesses into well-defined lows and highs. In effect, the repeater sharpens the differences between the signal's high and low points, making it easier for other hardware to interpret the bits.

So much for bits. Let's talk about qubits.

Quantum networks

Bits are sturdy, but qubits are very fragile. As we've already seen, measuring a qubit destroys the qubit's superposition. In addition, qubits suffer from **decoherence** – the natural loss of information due to inevitable interactions with the environment. As qubits travel from one point to another, they interact with any particles that they encounter. These interactions make the qubits behave more and more like plain old bits.

In today's quantum computers, decoherence is a big problem. *Chapter 3* included a small example in which decoherence makes a difference, but most of this book simply ignores the issue. After all, this book is about algorithms, not hardware, but in practice, physicists and engineers always factor decoherence into their quantum computing models.

Managing decoherence within a single quantum circuit is difficult enough, but limiting decoherence between two circuits in two different quantum computers is a monumental problem. In *Figure 6.2*, we imagine a qubit traveling from its source to a destination. (Don't worry. The numbers **0.5011** and **0.8654** in *Figure 6.2* aren't important. They're arbitrary values whose squares add up to 1.)

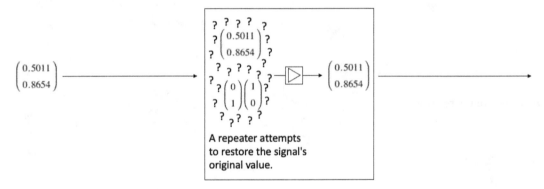

Figure 6.2 – A qubit's questionable journey

The qubit starts off in state $0.5011\,|0\rangle + 0.8654\,|1\rangle$. After traveling partway to its destination, the qubit might interact with something in its environment. A repeater receives a qubit that may or may not be in its original $0.5011\,|0\rangle + 0.8654\,|1\rangle$ state. In either case, the repeater should send a $0.5011\,|0\rangle + 0.8654\,|1\rangle$ qubit onward. But the properties of qubits present serious roadblocks. When a qubit collapses into either the $|0\rangle$ or $|1\rangle$ state, there's no turning back. You can't say, "Oops! This qubit shouldn't have been measured. I'll put it back into whatever superposition it had before the measurement." In this simple scenario, there's no way for the repeater to "know" what the qubit's starting state was.

In *Figure 6.3*, we imagine another situation. We assume that we can reliably transmit a qubit along a one-meter path, but our target destination is two meters from the source. A repeater receives an incoming qubit and sends out a fresh copy for the next one-meter leg of the trip.

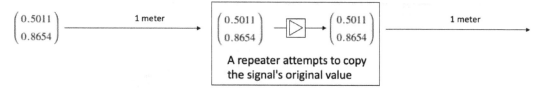

Figure 6.3 – Another failed attempt to transmit a qubit

Unfortunately, the strategy described in *Figure 6.3* is impossible. The No-Cloning theorem from *Chapter 5* says that a qubit can't be copied. Besides, if a qubit is already in a certain state, there's no such thing as having a "better copy" of that same state. So much for this idea!

The schemes proposed in *Figures 6.2* and *6.3* don't work. To send quantum information from one device to another, we need a different plan. And fortunately, we have one. To find out what it is, just keep reading.

Teleporting a qubit

This section describes a rudimentary scenario in which the state of a receiver's qubit becomes identical to the *original state* of a sender's qubit. We call this **quantum teleportation**, or **teleportation** for short. In our description of teleportation, the phrase *original state* is important. After all, a qubit can't be cloned. The sender's qubit may start in state $0.5011 |0\rangle + 0.8654 |1\rangle$, but it ends up in state $|0\rangle$ or state $|1\rangle$. In the meantime, the receiver's qubit goes into state $0.5011 |0\rangle + 0.8654 |1\rangle$.

Here's how teleportation works.

A device sits halfway between a sender named Alice and a receiver named Bob. We call this device a **repeater**, but it's nothing like the repeaters in our classical networking examples. The device is closer to both the sender and the receiver than they are to each other. In fact, the device can reliably send qubits to both the sender and the receiver. These qubits are entangled with one another, and that makes teleportation possible. *Figure 6.4* gives you the basic idea.

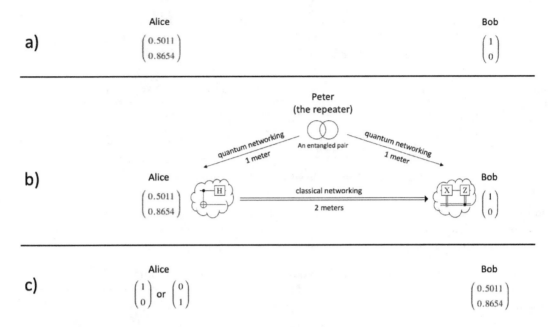

Figure 6.4 – Teleporting a qubit

Here's what happens in *Figure 6.4*:

a) At first, Alice's qubit is in state $0.5011 |0\rangle + 0.8654 |1\rangle$ and Bob's is in state $|0\rangle$.

b) Next, Peter takes advantage of his relative proximity to both Alice and Bob by sending each of them one of two entangled qubits.

After receiving Peter's qubit, Alice performs some quantum operations. Then, Alice uses an ordinary classical network to send information to Bob. Notice! The relatively long-distance communication between Alice and Bob takes place over a tried-and-true classical channel. That's nice!

c) Finally, Bob uses Alice's information to perform some quantum operations. Through some wizardry that's described in the next section, Bob's qubit ends up in state $0.5011 |0\rangle + 0.8654 |1\rangle$. (Alice's qubit ends up in state $|0\rangle$ or state $|1\rangle$.)

In the previous paragraph, we refer to some "*wizardry*" that puts Bob's qubit in the same state as Alice's original qubit state. The next section explains how that wizardry works.

Quantum operations for teleportation

To understand teleportation, you have to combine a few building blocks. We introduced some of the blocks in previous chapters. But a new block – my so-called "wizardry" step – requires some careful consideration. All in all, the idea is this:

Entangle Alice's original qubit and Bob's final qubit in a way that forces these qubits to have the same numbers in their amplitudes.

Figure 6.5 sets the stage for our discussion of teleportation:

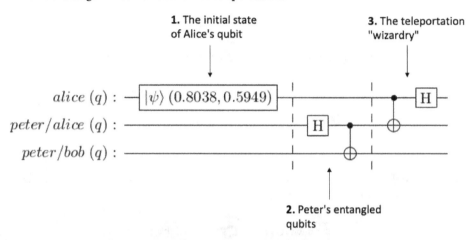

Figure 6.5 – How teleportation happens

Figure 6.5 is divided into three parts. Here's what happens in each part:

1. Alice's qubit begins in some arbitrary initial state. In *Figure 6.5*, the initial state is $0.8038\,|0\rangle\ +\ 0.5949\,|1\rangle$, but this initial state could be any valid qubit state. Instead of the numbers 0.8038 and 0.5949, let's use the letters α and β. Then, the three qubits in *Figure 6.5* start out looking like this:

$$|00\rangle(\,\alpha|0\rangle\ +\ \beta|1\rangle\,)$$

2. Peter creates an entangled pair in the usual way. He applies a Hadamard gate followed by a CNOT gate. Now we have the following formula for the three qubits in *Figure 6.5*:

$$\frac{1}{\sqrt{2}}(\,|00\rangle\ +\ |11\rangle\,)(\,\alpha|0\rangle\ +\ \beta|1\rangle\,)$$

    If you cross-multiply this formula, you get

$$\frac{1}{\sqrt{2}}(\,\alpha|000\rangle\ +\ \beta|001\rangle\ +\ \alpha|110\rangle\ +\ \beta|111\rangle\,)$$

    Peter sends one of the entangled pair's qubits to Alice and the other to Bob.

3. Alice applies a CNOT to the two qubits that she has. *Figure 6.6* illustrates the effect of this operation on the circuit's top and middle qubits:

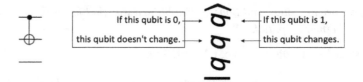

Figure 6.6 – The effect of applying CNOT to the upper two qubits

The result is to change the middle value in two of the formula's four kets:

$$\frac{1}{\sqrt{2}}(\ \alpha|000\rangle\ +\beta|011\rangle\ +\ \alpha|110\rangle\ +\beta|101\rangle\ )$$

Next, Alice applies a Hadamard gate to her original qubit. *Figure 6.7* illustrates this operation's effect on the circuit's top and middle qubits:

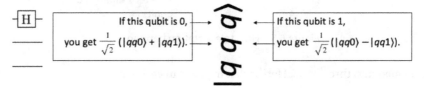

Figure 6.7 – The effect of applying H to the uppermost qubit

The result is to replace each ket with two kets:

$$\frac{1}{2}(\alpha(|000\rangle\ +\ |001\rangle\ )\ +\ \beta(|010\rangle\ -\ |011\rangle\ )\ +\ \alpha(|110\rangle\ +\ |111\rangle\ )\ +\ \beta(|100\rangle\ -\ |101\rangle\ ))$$

In this final result, each ket has three qubits. Alice has the middle and rightmost of these qubits. On a whim, let's factor out these two qubits belonging to Alice:

$$\frac{1}{2}[\ (\ \alpha|0\rangle + \beta|1\rangle\ )|00\rangle\ +$$
$$(\ \alpha|0\rangle - \beta|1\rangle\ )|01\rangle\ +$$
$$(\ \beta|0\rangle + \alpha|1\rangle\ )|10\rangle\ +$$
$$(-\beta|0\rangle + \alpha|1\rangle\ )|11\rangle\quad]$$

Notice that no matter what values Alice's two qubits have, *the state of Bob's qubit contains the numbers* α *and* β.

Bob's qubit is similar to Alice's original qubit. The only differences are these:

- In Bob's qubit, the roles of α and β may be reversed

- Bob's qubit may contain an extra minus sign

Fortunately, we have a quantum operation that reverses the roles of α and β. It's our old friend, the X gate:

$$X \begin{pmatrix} \alpha \\ \beta \end{pmatrix} = \begin{pmatrix} 0 & 1 \\ 1 & 0 \end{pmatrix} \begin{pmatrix} \alpha \\ \beta \end{pmatrix} = \begin{pmatrix} \beta \\ \alpha \end{pmatrix}$$

We also have an operation that changes a minus sign into a plus sign. It's called a **Z gate**:

$$Z \begin{pmatrix} \alpha \\ \beta \end{pmatrix} = \begin{pmatrix} 1 & 0 \\ 0 & -1 \end{pmatrix} \begin{pmatrix} \alpha \\ \beta \end{pmatrix} = \begin{pmatrix} \alpha \\ -\beta \end{pmatrix}$$

Even better, we know exactly when to apply each of these operations. Alice simply measures her two qubits:

- If Alice gets $|00\rangle$, then Bob has $\alpha|0\rangle + \beta|1\rangle$, which is identical to Alice's original qubit.

- If Alice gets $|01\rangle$, then Bob has $\alpha|0\rangle - \beta|1\rangle$. To turn this into $\alpha|0> + \beta|1>$, we apply a Z gate.

- If Alice gets $|10\rangle$, then Bob has $\beta|0\rangle + \alpha|1\rangle$. To turn this into $\alpha|0> + \beta|1>$, we apply an X gate.

- If Alice gets $|11\rangle$, then Bob has $-\beta|0\rangle + \alpha|1\rangle$. To turn this into $\alpha|0> + \beta|1>$, we apply an X gate and a Z gate.

The complete circuit, with Alice's measurements and Bob's extra gates, is shown in *Figure 6.8*.

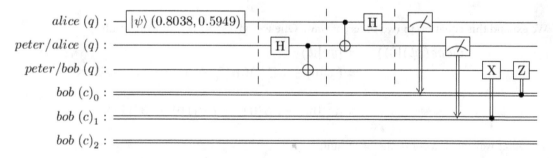

Figure 6.8 – The circuitry of quantum teleportation

In the fourth and final part of *Figure 6.8*, Alice measures her two qubits and gets plain old bits. She sends these bits to Bob along a classical network. Bob uses these bits to decide which of his X and Z gates (if any) to apply. The vertical lines below the X and Z gates indicate **conditional operations**. A conditional operation tells Qiskit to apply a quantum gate if and only if some classical bit has a particular value. A conditional operation is similar to a CNOT, except for the following:

- The control is a classical bit, not a quantum qubit.
- The target gate isn't necessarily an X gate.
- The bit value that makes the quantum gate fire isn't necessarily 1. It can be either 0 or 1.

As we'll see in this chapter's *Coding the teleportation circuitry* section, those conditional operations present a challenge for IBM's quantum hardware.

Teleportation versus cloning

In *Chapter 5*, we showed that a qubit can't be cloned. That is, if you start with a qubit in a certain state, you can't use that qubit to make a second qubit that's in the very same state. But in this chapter, Alice starts out with a qubit in a certain state, and Bob ends up with a qubit in that same state. What's going on here?

The trick in teleportation is that Alice measures her qubit. She destroys her qubit's superposition and sacrifices the qubit's original state to Bob. You can see this measurement in the fourth-from-last gate in *Figure 6.8*. Without a destructive act on Alice's qubit, the transfer of state would be impossible.

To bring this concept home, let's revisit the equations in *Chapter 5* and find out where that chapter's No-Cloning Theorem turns into this chapter's Yes-Teleporting-Possibility Theorem.

Let's start with a capsule summary of the reasoning in *Chapter 5*. We start by wrongly assuming the existence of a clone circuit. The circuit's equation is

$$Clone(|\psi\rangle|0\rangle) = |\psi\rangle|\psi\rangle, \quad \text{where } |\psi\rangle = \alpha|0\rangle + \beta|1\rangle$$

We expand this equation in two different ways. One way gives us the following result:

$$Clone(|\psi\rangle|0\rangle) = |\psi\rangle|\psi\rangle$$
$$= (\alpha|0\rangle + \beta|1\rangle)(\alpha|0\rangle + \beta|1\rangle)$$
$$\cdots$$
$$= \alpha^2|00\rangle + \alpha\beta|01\rangle + \alpha\beta|10\rangle + \beta^2|11\rangle$$

The other way yields a contradictory result:

$$Clone(|\psi\rangle|0\rangle) = Clone(\alpha|0\rangle|0\rangle + \beta|1\rangle|0\rangle)$$
$$\cdots$$
$$= \alpha|00\rangle + \beta|11\rangle$$

So, our assumption that the clone circuit exists is false.

When we try the same trick with a teleportation circuit, we don't get a contradiction. Here's why:

$$Teleport(|\psi\rangle|0\rangle) \ = \ |0\rangle|\psi\rangle, \quad \text{where } |\psi\rangle \ = \ \alpha|0\rangle + \beta|1\rangle$$

The teleport circuit's end result is $|0\rangle|\psi\rangle$ instead of $|\psi\rangle|\psi\rangle$. Now, let's expand the equation in two different ways:

$$
\begin{aligned}
Teleport(|\psi\rangle|0\rangle) \ &= |0\rangle|\psi\rangle \\
&= |0\rangle(\alpha|0\rangle + \beta|1\rangle) \\
&= |0\rangle\alpha|0\rangle + |0\rangle\beta|1\rangle \\
&= \alpha|00\rangle + \beta|01\rangle
\end{aligned}
$$

$$
\begin{aligned}
Teleport(|\psi\rangle|0\rangle) \ &= Teleport((\alpha|0\rangle + \beta|1\rangle)\,|0\rangle) \\
&= Teleport(\alpha|0\rangle|0\rangle + \beta|1\rangle|0\rangle) \\
&= \alpha\,Teleport(|0\rangle|0\rangle) + \beta\,Teleport(|1\rangle|0\rangle) \\
&= \alpha|0\rangle|0\rangle + \beta|0\rangle|1\rangle \\
&= \alpha|00\rangle + \beta|01\rangle
\end{aligned}
$$

The end result is the same no matter how we expand the teleport equation. Unlike the clone calculation in *Chapter 5*, we don't have a contradiction.

We should note that the equations in this section don't prove that teleportation is possible. Proving that it's possible was the work of the previous section – the section titled *Teleporting a qubit*. This section's calculations only assure us that whatever reasoning doomed cloning in *Chapter 5* doesn't come back to haunt us when we do teleportation here in *Chapter 6*.

Coding the teleportation circuitry

In this section, we will provide code to teleport qubits and test the results. We've divided the code into three parts:

- Creating registers
- Adding gates to the registers
- Running the quantum circuit

Creating registers

We start with the imports:

```
from qiskit import QuantumCircuit, QuantumRegister, \
    ClassicalRegister, Aer
import random
import numpy as np
from qiskit.result import marginal_counts
```

In this chapter, the only new import is `marginal_counts`. We will describe that function later in this section.

The following function defines a circuit:

```
def create_registers():
    alice_q = QuantumRegister(1, 'alice (q)')
    peter_alice_q = \
        QuantumRegister(1, 'peter/alice (q)')
    peter_bob_q = QuantumRegister(1, 'peter/bob (q)')
    bob_c = ClassicalRegister(3, 'bob (c)')
    circ = QuantumCircuit(alice_q, peter_alice_q,
                          peter_bob_q, bob_c)
    return circ
```

A call to this code creates three qubits and three classical bits, as shown in *Figure 6.9*.

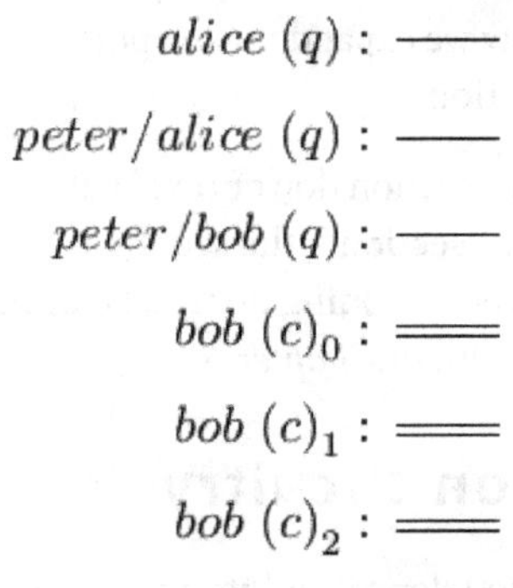

$$alice\ (q): \longrightarrow$$

$$peter/alice\ (q): \longrightarrow$$

$$peter/bob\ (q): \longrightarrow$$

$$bob\ (c)_0: =\!=\!=$$

$$bob\ (c)_1: =\!=\!=$$

$$bob\ (c)_2: =\!=\!=$$

Figure 6.9 – The registers required for teleportation

In *Figure 6.9*, Alice starts with one qubit, the repeater sends qubits to both Alice and Bob, and Bob receives all three of the algorithm's measurements.

Adding gates to the registers

The next piece of code randomly decides on an initial state for Alice's qubit:

```
def generate_amplitudes():
```

```
alpha = np.sqrt(random.uniform(0, 1))
beta = np.sqrt(1 - alpha**2)
return alpha, beta
```

We need a function that adds gates to the registers in *Figure 6.9*. Here's the code for that function:

```
def add_gates(circ, alpha, beta):
    circ.initialize([alpha, beta], 0)
    circ.barrier()
    circ.h(1)
    circ.cnot(1, 2)
    circ.barrier()
    circ.cnot(0, 1)
    circ.h(0)
    circ.barrier()
    circ.measure(0, 0)
    circ.measure(1, 1)
    with circ.if_test((1, 1)):
        circ.x(2)
    with circ.if_test((0, 1)):
        circ.z(2)
    circ.measure(2, 2)
    return circ
```

Figure 6.10 shows a circuit created by a call to the add_gates function. This circuit is almost identical to the circuit in *Figure 6.8*. But at the end of *Figure 6.10*, we have an additional measurement gate. Our strategy is to count the number of zeros and ones that come from this measurement and check to make sure that it's what we would expect from the original state of Alice's qubit.

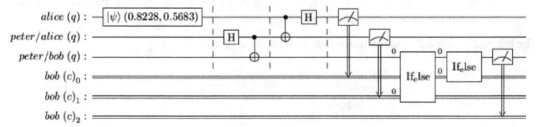

Figure 6.10 – A circuit to test teleportation

A few statements in the add_gates function are worth noting:

- Qiskit's initialize function starts a qubit in some particular state. It can override the default initial state of |0⟩. In this code, the circ.initialize([alpha, beta], 0) statement puts the qubit with index 0 in the state alpha|0⟩ + beta|1⟩.

When Qiskit draws the circuit, we see a rectangle containing those `alpha` and `beta` values. For example, when I created the circuit in *Figure 6.10*, `alpha` was 0.8228, and `beta` was 0.5683.

> **Important note**
>
> In *Figure 6.10*, the circuit's `draw` function shows only four digits of accuracy in the initialization of Alice's qubit. In this example, the randomly generated values of `alpha` and `beta` aren't exactly 0.8228 and 0.5683. In fact, if you sum up the squares of 0.8228 and 0.5683, you don't get 1. Instead, you get 0.99996473.

- Inside the `add_gates` function, the `with...if_test` statements create conditional operations. In *Figure 6.10*, these operations appear near the end of the circuit. In general, `with...if_test` gets a classical register's value and uses that value to decide whether or not it will apply a particular quantum gate.

  For example, our `add_gates` function contains the following code:

  ```
  with circ.if_test((0, 1)):
      circ.z(2)
  ```

  This code applies a Z gate to the quantum qubit with index 2 if and only if the classical bit with index 0 has the value 1. The other `with...if_test` code in the `add_gates` function behaves in a similar fashion. (See *Figure 6.10*.)

  Concerning the `if_test` method, some cautions are in order. In our `add_gates` function, each `if_test` call has only one parameter, and that parameter is a tuple. That's why the `circ.if_test(0, 1)` call, without double parentheses, would be invalid.

  Also, you may be tempted (as I was) to forgo Qiskit's `with...if_test` statements and use Python's tried-and-true `if` statements instead. That won't work. Python would evaluate the `if` statement while it's constructing the circuit, not while the quantum computer is running the circuit. Try it yourself and see what happens!

The following code calls the functions that we've defined so far:

```
alpha, beta = generate_amplitudes()
circ = create_registers()
circ = add_gates(circ, alpha, beta)
display(circ.draw('latex', cregbundle=False))
```

In this code's `circ.draw` call, `cregbundle=False` tells Qiskit to draw a separate line for each bit in Bob's classical register. In my opinion, this helps clarify the roles of the bits in the teleportation algorithm.

Running the quantum circuit

In the last chunk of code, we finally run the teleportation test:

```
device = Aer.get_backend("qasm_simulator")

shots = 1000
job = device.run(circ, shots=shots)
print(job.job_id())

result = job.result()
counts = result.get_counts(circ)
counts_m = marginal_counts(counts, [2])
number_of_0s = counts_m.get('0')
number_of_1s = counts_m.get('1')
alpha = np.sqrt(number_of_0s / shots)
beta = np.sqrt(number_of_1s / shots)
print("|\u03C8\u27E9 ({:.4f}, {:.4f})".format(alpha, beta))
```

> **Important note**
>
> Conditional gates that go from classical bits to quantum registers present some special implementation problems, so the stewards of Qiskit introduced this feature slowly. The `if_test` method didn't become an official part of Qiskit until I double-checked this chapter in 2023. As we headed toward publication, `if_test` wasn't fully merged into the Qiskit ecosystem. To test this program, I had to switch from `execute` to calling `run`. By the time you read this book, this wrinkle should be ironed out.

Our circuit has three measurement gates, but we're interested in the counts from only one of them. So, in this code, Qiskit's `marginal_counts` function sifts the counts from the qubit with index 2 and ignores the others.

In the last five statements, the code uses the counts to find the state of Bob's qubit. When I ran this code on the circuit from *Figure 6.10*, I got the following output:

```
|ψ⟩ (0.8343, 0.5514)
```

Those numbers are pretty close to the alpha and beta values in *Figure 6.10*. Other runs with other randomly generated alpha and beta values, were also encouraging. In a serious research situation, we'd do further calculations to make sure that the results of our runs were statistically significant, but statistical analysis is a topic in and of itself. In the absence of any rigorous statistical analysis, I'm happy with the results.

Summary

Transmitting qubit states from one place to another is vastly different from transmitting classical bits. For a qubit, the tiniest interaction with anything at all causes decoherence, which collapses the qubit's state. So, the transmission of a qubit over a significant distance isn't feasible.

But you can teleport a qubit's state. With teleportation, a repeater stands halfway between a sender and a receiver. After creating an entangled pair, the repeater transmits one of the qubits to the sender and the other to the receiver. This cuts the qubit travel distance in half.

The state of the sender's qubit doesn't survive the teleportation process because the sender measures their qubit and sends the resulting bit to the receiver. And sending a classical bit over a considerable distance presents no difficulties. We've done so for several decades.

So far, in this book, we've described ways in which quantum computers manipulate qubits, but we haven't used qubits to answer questions or solve problems. In the next chapter, we will define a tiny problem and show that quantum computers excel in finding solutions to that problem.

Questions

1. This chapter's *Quantum operations for teleportation* section presents the equations that show why teleportation works. What happens in these equations if the initial state of Alice's qubit is $|0\rangle$? What happens if the initial state is $|1\rangle$?

2. In *Figure 6.10*, Alice's qubit starts in state $0.8228\,|0\rangle + 0.5683\,|1\rangle$. Modify this chapter's generate_amplitudes function as follows:

```
def generate_amplitudes():
    alpha = 0.8228
    beta = 0.5683
    return alpha, beta
```

With this version of generate_amplitudes, what happens when you call the add_gates function? Why?

3. Write Qiskit code to create a circuit of the following kind:

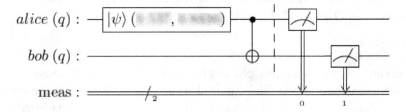

I've blurred out the initial amplitudes of Alice's qubit because you should generate those values randomly.

Run this circuit several times to convince yourself that Bob's qubit ends up in the same state as Alice's qubit. To further convince yourself, calculate the state of Bob's qubit using matrix arithmetic.

With Bob's qubit in the same state as Alice's qubit, why don't we claim that this circuit teleports Alice's qubit?

4. Look again at *Figure 6.5*. Just before part **3**, the qubits are in the state $\frac{1}{\sqrt{2}}(\,|00\rangle + |11\rangle\,)(\,\alpha|0\rangle + \beta|1\rangle\,)$ where $\alpha = 0.8038$ and $\beta = 0.5949$. The gates in part **3** apply $(\,I \otimes I \otimes H\,)(\,I \otimes CNOT\,)$ to that state. Do the matrix arithmetic to show that, after part **3**, we have the following state:

$$\frac{1}{2}(\,\alpha(|000\rangle + |001\rangle) + \beta(|010\rangle - |011\rangle) + \alpha(|110\rangle + |111\rangle) + \beta(|100\rangle - |101\rangle))$$

5. In *Figure 6.4*, the maximum distance for reliable qubit transmission is one meter, but Alice and Bob are two meters apart. Modify the code in this chapter's *Coding the teleportation circuitry* section so that the sender and receiver are three meters apart. Have Alice teleport her qubit's state to Bob, who uses teleportation to forward the qubit's state to Carol.

Further reading

[1]. https://datareportal.com/reports/digital-2022-october-global-statshot (page 40)

Part 3
Quantum Computing
Algorithms

In this part, we will solve computational problems using quantum computers. For each problem, the quantum algorithm requires fewer steps than the corresponding classical algorithm. Deutsch's algorithm finds information about all of a function's outputs by evaluating the function only once. Grover's algorithm searches through a list without examining each of the list's items individually. Shor's algorithm (unfortunately) cracks the widely-used RSA encryption scheme.

This part has the following chapters:

- *Chapter 7, Deutsch's Algorithm*
- *Chapter 8, Grover's Algorithm*
- *Chapter 9, Shor's Algorithm*

7
Deutsch's Algorithm

Classical computers solve problems. If they didn't solve any problems, we'd call them "classical waste."

So, what about quantum computers? Are there any problems that quantum computers can solve? If you've read *Chapters 1* to *6*, you may be suspicious.

Chapters 1 to *4* describe qubits, but those chapters say nothing about problems you can solve with those qubits. *Chapters 5* and *6* show you how to move qubits around, but neither of those chapters involves a question-and-answer scenario. You can ask a classical computer whether 15 equals 3 times 5. Could you ask a quantum computer to do that?

In this chapter, we pose a question and a quantum computer provides an answer. The question concerns binary-valued functions, and the answer comes courtesy of something called **Deutsch's algorithm**.

When you read about Deutsch's algorithm, your first impression may be that this algorithm is a complete waste of resources. Here's why:

- The problem that Deutsch's algorithm solves is extremely simple. If you posed this problem to a chatbot, the bot's cynical answer might be `Don't be so lazy! You can solve this problem by hand!`.

- The problem that Deutsch's algorithm solves is contrived. As you travel through life, you may never have a need to solve this problem.

- The technique used in Deutsch's algorithm bears no resemblance to any step-by-step procedure that you'd apply in classical computing. I can describe *what* Deutsch's algorithm is in just a few lines of code. But I need several sections to explain *why* Deutsch's algorithm works.

In a way, Deutsch's algorithm consumes more resources and produces less useful results than a typical `Hello world` example in an elementary programming book. But, as a proof of concept for quantum computing, Deutsch's algorithm is an enormous leap forward. This chapter tells you all about it.

Topics include the following:

- Describing Deutsch's problem

- Solving Deutsch's problem
- Deutsch's algorithm F.A.Q.
- Coding Deutsch's algorithm

Technical requirements

The code used in this chapter can be found online on GitHub at `https://github.com/PacktPublishing/Quantum-Computing-Algorithms`.

Describing Deutsch's problem

Deutsch's algorithm solves a problem concerning constant and balanced functions. So, as you start this chapter, I want to make sure that you're familiar with the terms *algorithm*, *function*, *constant function*, and *balanced function*. This section covers those essential ideas.

Algorithms

You may have seen friendly signs that display the equivalent of "Welcome!" in many languages. *Figure 7.1* does the same kind of thing for a particular sequence of instructions:

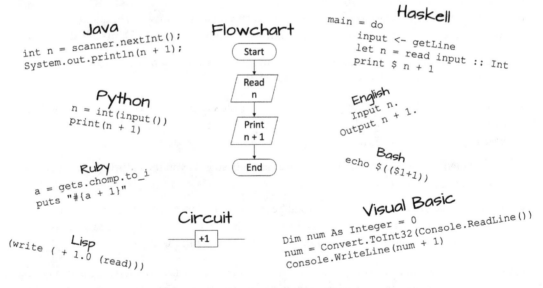

Figure 7.1 – Expressing the steps that add 1 to a number

Each part of *Figure 7.1* inputs a number and outputs that number plus 1. The parts look different, but they all describe the same sequence of steps. This sequence of steps is an example of an algorithm.

> **Definition**
> An **algorithm** is a sequence of steps.

You can change the way you describe a particular sequence of steps. But when you do, you still have the same algorithm.

If you have two different sequences of steps, you have two different algorithms. One algorithm for traveling from New York to London is to take a 7-day cruise. A different algorithm involves flying from New York to London with a stopover in Paris.

Of course, some algorithms have wiggle room. When you replace butter with margarine in the recipe for tuna casserole, is that a new algorithm or a slight variation on an existing algorithm? It's up to you to decide.

Many authors will add to my definition of "algorithm" by insisting that an algorithm solves a particular problem and delivers a result after executing a finite number of steps. I've seen algorithms with neither of these characteristics, but I don't want to start an argument about it.

Functions

Here's more terminology:

> **Definition**
> A **function** is a rule that associates a particular output value to every one of its allowable input values.

For example, the rule $f(x) = x + 10$ assigns the output value 11 to the input value 1, assigns the output value 12 to the input value 2, and so on. With the function f defined this way, we can write the following:

$$f(1) = 11$$
$$f(2) = 12$$
$$\cdots \text{ and so on.}$$

It's worth noting that a function isn't the same as a mathematical formula. Some functions have nothing to do with formulas. Consider the following bestFriendOf function:

$$\text{bestFriendOf(Joe)} = \text{Jane}$$
$$\text{bestFriendOf(Sue)} = \text{Mary}$$
$$\text{bestFriendOf(Alan)} = \text{Ed}$$

This function has no particular rhyme or reason. It's just a rule that assigns an output value (such as Jane, Mary, or Ed) to each of its allowable input values (namely Joe, Sue, and Alan).

Constant and balanced functions

Speaking of functions, consider the following:

Outputs	Name	Formula
$f(0) = 0$ $f(1) = 0$	The *Zero* function	$f(x) = 0$

This function's only allowable input values are 0 and 1. With both input values, the function outputs the value 0. We call this a **constant** function because, no matter which input value we give the function, its output value doesn't change. Here's another constant function:

Outputs	Name	Formula
$f(0) = 1$ $f(1) = 1$	The *One* function	$f(x) = 1$

No matter which input value we give this function, its output value is always 1.

Both the *Zero* and *One* functions are simple binary functions. I call them **simple binary** functions because their only allowable inputs are 0 and 1, and their only allowable outputs are 0 and 1.

Once we decide to create a simple binary function, we have two possibilities besides the *Zero* and *One* functions. Here's a possibility:

Outputs	Name	Formula
$f(0) = 0$ $f(1) = 1$	The *Same_as* function	$f(x) = x$

And here's the other possibility:

Outputs	Name	Formula
$f(0) = 1$ $f(1) = 0$	The *Opposite_of* function	$f(x) = \neg x$

In the formula for the *Opposite_of* function, we represent "opposite" with the $\neg$ symbol. If x is 0, then $\neg x$ is 1. And if x is 1, then $\neg x$ is 0.

Neither the *Same_as* nor the *Opposite_of* function is constant. Instead, these functions are balanced. A **balanced** binary function is a function that outputs 0 and 1 in equal amounts. The *Same_as* function outputs one 0 and one 1. The *Opposite_of* function also outputs one 0 and one 1.

We have four simple binary functions, and each one of them is either constant or balanced. In this chapter, we don't consider functions that aren't binary. But if we did, we'd have functions that are neither constant nor balanced. For example, the following function has two 0 outputs but only one 1 output:

$$f(0) = 0$$
$$f(1) = 1$$
$$f(2) = 0$$

This function isn't binary, and it's neither constant nor balanced.

Deutsch's problem

A long time ago, I bought action figures as gifts for small children. Five figures portrayed Parker Porcupine, a favorite among kids that year. Another five figures were of Zeke the Zebra, a less popular character in an unimaginative cartoon show. (What could I do? The store had only five Parker Porcupines in stock.) I wrapped all ten gifts and gave eight of them to my co-workers for their kids.

Now, I had two wrapped action figures. With two kids of my own, I needed to know something about those wrapped figures. Were they two Parker Porcupines? That would be fine. Were they two Zeke the Zebras? That wouldn't be so bad. But if one was Parker Porcupine and the other was Zeke the Zebra, I'd be in big trouble. One kid would gloat over getting the better gift, and the other would whine about being short-changed.

What I had was a function from wrapped gifts to cartoon characters. If the function was constant, my kids would be happy. But if the function was balanced, I'd be in trouble.

How could I find out what I wanted to know about these gifts? Reopening one of them would give me no useful information. Alas, I had to open both of them.

Suppose that a friend of yours—call her Nora—hands you a tiny component. It's part of a quantum circuit. She tells you that this component implements one of the four simple binary functions. She doesn't tell you whether the component implements a constant or a balanced function. How can you find out?

You add the component to your own circuitry and input the value 0. As a result, the circuit outputs 0. So, you know that Nora's component implements either the *Zero* function or the balanced *Same_as* function. But you don't know which of these functions it implements. To know for sure whether the component's function is constant or balanced, you have to feed the circuit a second input.

Here's the classical algorithm for testing the component:

```
Input 0 and store the output in the variable a.
Input 1 and store the output in the variable b.
If a equals b,
    the function is constant;
else,
    the function is balanced.
```

In the classical world, you feed two inputs to the component. In other words, you evaluate the component's function two times.

But the quantum world is different. A quantum computer can decide whether the function is constant or balanced after evaluating the function only once. That's Deutsch's algorithm.

The following sections tell you how Deutsch's algorithm works.

Solving Deutsch's problem

The circuit that solves Deutsch's problem has very few gates. It doesn't look very complicated. But to understand why the circuit works, you have to understand some important concepts. This section covers those critical concepts.

Phase kickback

In *Chapter 4*, we introduced the **controlled NOT (CNOT)** gate with its control and target qubits. In a typical scenario, the control qubit tells the target qubit what to do, and the target qubit obeys willingly. But what about a non-typical scenario? Is there such a thing as a disobedient target qubit? To find out, let's do an experiment.

Run the following code:

```
from qiskit import QuantumCircuit
from qiskit.quantum_info import Statevector
from qiskit.visualization \
    import plot_bloch_multivector

circ = QuantumCircuit(2)
circ.x(1)
circ.h(0)
circ.h(1)
display(circ.draw('latex'))

state = Statevector(circ)
display(plot_bloch_multivector(state, reverse_bits=True))
```

Figure 7.2 shows this code's output:

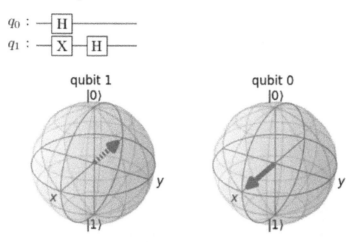

Figure 7.2 – Applying H gates to |1⟩ and |0⟩ qubits

> **Important note**
>
> As in previous chapters, we've doctored some of Qiskit's output to make the display clearer. In your Qiskit notebook, the arrows inside both spheres have solid lines. But in *Figure 7.2*, we touched up Qiskit's Bloch sphere diagrams so that any arrow pointing to the back of the sphere is a dotted line. In addition, we surreptitiously added `figsize=(3, 3), font_size=12` to the `plot_bloch_multivector` call. This adjusted the proportions in *Figure 7.2*.

In *Figure 7.2*, the H gates do what we expect them to do:

- With q_0, applying H to |0⟩ gives us $\frac{1}{\sqrt{2}}$(|0⟩ + |1⟩)), also known as |+⟩

- With q_1, applying H to |1⟩ gives us $\frac{1}{\sqrt{2}}$(|0⟩ + |1⟩)), also known as |–⟩

This is our first example in which `plot_bloch_multivector` draws more than one qubit. If we omitted the `reverse_bits` parameter from the `plot_bloch_multivector` call, qubit 0 would appear on the left, with qubit 1 on the right.

Let's add a CNOT gate to the end of this circuit:

```
circ.cnot(0, 1)
display(circ.draw('latex'))
state = Statevector(circ)
display(plot_bloch_multivector(state, reverse_bits=True))
```

When you run this cell's code, you get the output shown in *Figure 7.3*:

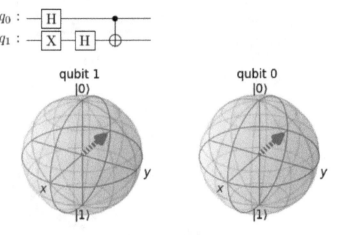

Figure 7.3 – Applying CNOT to the qubits in Figure 7.2

Between *Figures 7.2* and *7.3*, we added a CNOT gate. An informal description of a CNOT gate says, "CNOT applies NOT to the target qubit if and only if the control qubit is 1." So, between *Figures 7.2* and *7.3*, we expect the control qubit, q_0, to stay as it is, and the target qubit, q_1, to change in some way.

But look at the Bloch spheres in *Figure 7.3*. Between *Figures 7.2* and *7.3*, the state of the control qubit, q_0, changes, and the value of the target qubit, q_1, doesn't change! This strange phenomenon is called **phase kickback**. When the control qubit tells the target qubit to change, that change bounces off the target and lands back on the control qubit.

If you don't believe what you see in *Figures 7.2* and *7.3*, you can do the math. At the end of the circuit in *Figure 7.2*, $q_1 \otimes q_0$ is shown as follows:

$$|-\rangle \otimes |+\rangle = \frac{1}{\sqrt{2}}\begin{pmatrix} 1 \\ -1 \end{pmatrix} \otimes \frac{1}{\sqrt{2}}\begin{pmatrix} 1 \\ 1 \end{pmatrix} = \frac{1}{2}\begin{pmatrix} 1 \\ 1 \\ -1 \\ -1 \end{pmatrix}$$

After applying CNOT in *Figure 7.3*, you get the following:

$$\frac{1}{2}\begin{pmatrix} 1 & 0 & 0 & 0 \\ 0 & 0 & 0 & 1 \\ 0 & 0 & 1 & 0 \\ 0 & 1 & 0 & 0 \end{pmatrix}\begin{pmatrix} 1 \\ 1 \\ -1 \\ -1 \end{pmatrix} = \frac{1}{2}\begin{pmatrix} 1 \\ -1 \\ -1 \\ 1 \end{pmatrix} = \frac{1}{\sqrt{2}}\begin{pmatrix} 1 \\ -1 \end{pmatrix} \otimes \frac{1}{\sqrt{2}}\begin{pmatrix} 1 \\ -1 \end{pmatrix} = |-\rangle \otimes |-\rangle$$

And there you have it! The q_0 qubit started off in the $|+\rangle$ state and ended up in the $|-\rangle$ state.

Phase kickback is an example of a more general principle. In quantum mechanics, you can't ignore the impact of an observer. If you start with a qubit that's in a state of superposition, and you measure that qubit, the qubit snaps out of superposition and goes into the $|0\rangle$ or $|1\rangle$ state. Observations aren't passive events. When you observe something, you change something.

In a similar fashion, a control qubit isn't a passive bystander in the application of a CNOT state. Under certain conditions, a control qubit's interaction with its target causes changes in the control qubit itself.

If you're mystified by phase kickback and you want to learn more, read the following subsection.

When does kickback kick in?

Phase kickback isn't limited to the circuit in *Figure 7.3*. Phase kickback takes place when your circuit satisfies the following three elements:

1. A multi-qubit gate with a control qubit and a target qubit

2. A target qubit that doesn't move along the Bloch sphere when the control operation is applied

3. Something for the target control to kick back

Look again at *Figure 7.2*. Both qubits' arrows lie along the x axis, but let's concentrate on the target qubit (qubit 1). When you apply an X gate to qubit 1, that qubit's position on the Bloch sphere doesn't change. If anything, you can imagine that the qubit's arrow spins in place. Having the target qubit stay put is one of the required elements for phase kickback.

Let's consider another example. We create a circuit as follows:

```
circ = QuantumCircuit(2)
circ.h(0)
circ.x(1)
```

With calls to `circ.draw` and `plot_bloch_multivector`, we get the output shown in *Figure 7.4*:

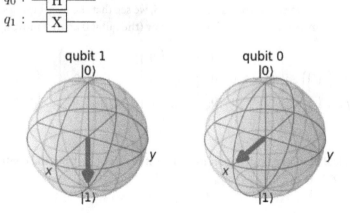

Figure 7.4 – A circuit and its qubit states

We can add a controlled Z gate to the end of this circuit, like so:

```
circ = QuantumCircuit(2)
circ.h(0)
circ.x(1)
circ.cz(0, 1) # Controlled Z gate
```

After adding the controlled Z gate, we get the spheres shown in *Figure 7.5*:

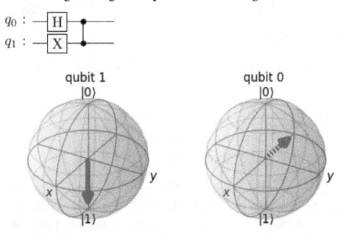

Figure 7.5 – Adding a Z gate to the circuit shown in Figure 7.4

In this code, the `circ.cz(0, 1)` statement creates a controlled Z gate with control qubit q_0 and target qubit q_1. So, this example has the first element that's required for phase kickback. But notice that, in *Figure 7.5*, the circuit diagram makes no distinction between the control qubit and the target qubit. For a Z gate, the distinction between the control qubit and the target qubit is insignificant, but that's another story.

Rotating the sphere around the z axis doesn't move the arrow of qubit 1. So, this example has the second element that's required for phase kickback. In *Figure 7.5*, we see that the target qubit's arrow (the qubit 1 arrow) doesn't move. Instead, the control qubit's arrow (the qubit 0 arrow) moves. Here's the math:

$$CZ\left(\begin{pmatrix} 0 \\ 1 \end{pmatrix} \otimes \frac{1}{\sqrt{2}}\begin{pmatrix} 1 \\ 1 \end{pmatrix}\right) = \begin{pmatrix} 1 & 0 & 0 & 0 \\ 0 & -1 & 0 & 0 \\ 0 & 0 & 1 & 0 \\ 0 & 0 & 0 & -1 \end{pmatrix} \frac{1}{\sqrt{2}}\begin{pmatrix} 0 \\ 0 \\ 1 \\ 1 \end{pmatrix} = \frac{1}{\sqrt{2}}\begin{pmatrix} 0 \\ 0 \\ 1 \\ -1 \end{pmatrix} = \begin{pmatrix} 0 \\ 1 \end{pmatrix} \otimes \frac{1}{\sqrt{2}}\begin{pmatrix} 1 \\ -1 \end{pmatrix}$$

Throughout this calculation, the target qubit maintains its value of $\begin{pmatrix} 0 \\ 1 \end{pmatrix}$, but the control qubit changes from $\frac{1}{\sqrt{2}}\begin{pmatrix} 1 \\ 1 \end{pmatrix}$ to $\frac{1}{\sqrt{2}}\begin{pmatrix} 1 \\ -1 \end{pmatrix}$.

To understand the third element in the criteria for phase kickback, we consider an example that doesn't have that element. We create a circuit as follows:

```
circ = QuantumCircuit(2)
circ.h([0, 1])
```

Before adding a CNOT gate, we have the output shown in *Figure 7.6*:

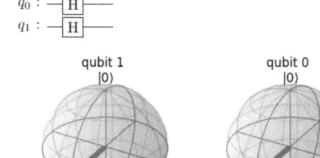

Figure 7.6 – Another circuit with qubit states

After adding a CNOT gate, we get the same Bloch spheres as in *Figure 7.6*. The math looks like this:

$$|+\rangle \otimes |+\rangle = \frac{1}{\sqrt{2}}\begin{pmatrix}1\\1\end{pmatrix} \otimes \frac{1}{\sqrt{2}}\begin{pmatrix}1\\1\end{pmatrix} = \frac{1}{2}\begin{pmatrix}1\\1\\1\\1\end{pmatrix}$$

$$CNOT(|+\rangle \otimes |+\rangle) = \frac{1}{2}\begin{pmatrix}1&0&0&0\\0&0&0&1\\0&0&1&0\\0&1&0&0\end{pmatrix}\begin{pmatrix}1\\1\\1\\1\end{pmatrix} = \frac{1}{2}\begin{pmatrix}1\\1\\1\\1\end{pmatrix} = |+\rangle \otimes |+\rangle$$

A CNOT gate always exchanges the values of the $|01\rangle$ and $|11\rangle$ amplitudes. But, in this circuit, the $|01\rangle$ and $|11\rangle$ amplitudes are equal. So, nothing changes when we apply the CNOT gate.

In the next section, we make use of the phase kickback phenomenon.

Detecting a CNOT gate

Let's return to your friend, Nora, from the *Deutsch's problem* section. Once again, Nora gives you a component that you can attach to your own circuitry. But this time, she says that the component contains either a CNOT gate or no gates at all (see *Figure 7.7*). It's your job to figure out if there's a CNOT gate inside this component:

Figure 7.7 – Does Nora's component contain a CNOT gate?

You attach the component to the end of the circuitry shown in *Figure 7.2*. After exiting from Nora's component, the state of q_0 is either $|+\rangle$ or $|-\rangle$. So, you add your own Hadamard and measurement gates to the end of your new circuit (see *Figure 7.8*):

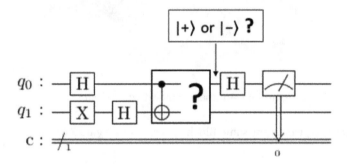

Figure 7.8 – Attaching Nora's circuit to your equipment

- A Hadamard gate turns $|+\rangle$ into $|0\rangle$. So, if the measurement yields 0, the component's output was $|+\rangle$. As in *Figure 7.2*, the component doesn't contain a CNOT gate.

- A Hadamard gate turns $|-\rangle$ into $|1\rangle$. So, if the measurement yields 1, the component's output was $|-\rangle$. As in *Figure 7.3*, the component contains a CNOT gate.

After running the circuit, you know whether or not Nora's component contains a CNOT gate.

In the next section, we explore the connection between binary functions and CNOT gates.

Embedding a function in quantum circuitry

In *Chapter 3, Math for Qubits and Quantum Gates*, we declared that quantum gate operations must be reversible. We can pass a qubit through any genuine quantum gate (almost any kind of gate except a measurement gate). If we know what kind of quantum gate we have, and we know the state of the qubit after it leaves the gate, we can calculate what the qubit's state was before that qubit entered the gate. In other words, we can undo any genuine quantum operation. *Figure 7.9* has two simple examples:

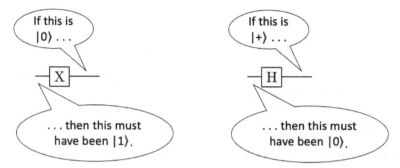

Figure 7.9 – The X and H gates are reversible

The same rule holds true when we combine gates to form a component. If we have a diagram of the gates inside the component, and we know the states of the qubits after they leave the component, we can calculate what the qubits' states were before they entered the component.

So, what about the *Zero* function that we defined in this chapter's *Constant and balanced functions* section? Does that function define a reversible operation? The answer is "no". *Figure 7.10* shows that the *Zero* function isn't reversible:

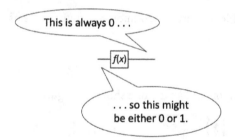

Figure 7.10 – Which input value gave you the 0 output value?

We know that the $f(x)$ component in *Figure 7.10* always outputs 0. But this information doesn't help us determine where that 0 came from. Was the component's input 0 or 1? We simply don't know.

To work with the *Zero* function in a quantum circuit, we have to embed that function inside a reversible component. The same holds true of the three other simple binary functions in the *Constant and balanced functions* section (the *One*, *Same_as*, and *Opposite_of* functions).

Let's think again about your friend, Nora, from the previous sections. Nora selects one of four simple binary functions. We'll name Nora's function $f(x)$. Nora wants to make reversible circuitry whose output includes $f(x)$. Nora makes several attempts. Her first attempt, the most obvious way to make a reversible circuit, doesn't work at all. But by examining a few of her failed attempts, we come to understand Nora's final success. Here's the story.

Nora's first attempt at making a reversible circuit

Can Nora use a single-qubit component like the one shown in *Figure 7.10*?

No. The component in *Figure 7.10* isn't reversible. There's not enough information coming out of the component to represent both the input value x and output value $f(x)$.

Nora's second attempt

Can Nora use a two-qubit component like the one shown in *Figure 7.11*?

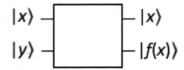

Figure 7.11 – Is this circuit reversible?

As with any two-qubit component, the component in *Figure 7.11* has two inputs and two outputs. The component's extra y input is called an **ancilla**. This y input plays no role in the definition of $f(x)$, but that's okay.

This component is not reversible because the component's output tells us nothing about the value of the y input.

Third attempt

Can Nora use a two-qubit component like the one shown in *Figure 7.12*?

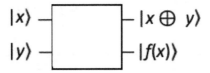

Figure 7.12 – What about this circuit? Is it reversible?

The $\oplus$ symbol in *Figure 7.12* is called **exclusive OR**, also known as **XOR**. The value of $x \oplus y$ is 1 if—and only if—either x or y (but not both) is 1.

Figure 7.13 shows you another way to think about the XOR operator:

x	y	x $\oplus$ y	
0	0	0	If x is 0, x $\oplus$ y is the same as y
0	1	1	
1	0	1	If x is 0, x $\oplus$ y is the same as ¬y
1	1	0	

Figure 7.13 – The XOR operation

It's as if $x \oplus y$ "wants" to be the same as y. The only thing that can stop this from happening is if $x = 1$.

Let's get back to the component shown in *Figure 7.12*. The component's output depends on all three of the values—x, y, and $f(x)$. Does that make this component reversible? Unfortunately, it doesn't. To understand why, consider the case in which $f(x)$ is the *Zero* function:

- If the component's inputs are both 0, then both of its outputs are 0

- If the component's inputs are both 1, then both of its outputs are 0

So, when $f(x)$ is the *Zero* function, and both outputs are 0, there's no way to tell whether the component's inputs were both 0 or both 1.

The component in *Figure 7.12* isn't reversible, but the use of XOR gives Nora an idea.

Fourth attempt

Can Nora use a two-qubit component like the one shown in *Figure 7.14*?

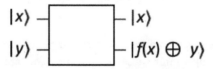

Figure 7.14 – Success at last!

Hooray! The component in *Figure 7.14* is reversible! Nora knows which of the simple four binary functions $f(x)$ is. She can examine the component's two outputs and use them to determine the component's two inputs. *Figure 7.15* shows you how:

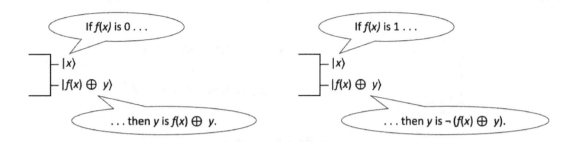

Figure 7.15 – Determining the inputs by examining the outputs

Nora has succeeded in creating a reversible component, but reversibility isn't enough. She also wants her component to output values of $f(x)$. Fortunately, the component in *Figure 7.14* can do that. When you input $|0\rangle$ for $|y\rangle$, the component's bottom output is $f(x) \oplus 0$, which is $f(x)$.

In the next section, we'll use X gates and CNOT gates to implement the component depicted in *Figure 7.14*, and we'll revisit the reversibility argument using matrix arithmetic.

Creating oracles

In ancient Greece, citizens posed difficult questions to the Oracle at Delphi. "Should I journey from Athens to Sparta? Would a sacrifice to the gods benefit my standing in the community? Which brand of Hemlock should I buy for Socrates?" The Oracle used its unfathomable powers to provide answers. No one knew how these powers worked. The Oracle was, in effect, a superhuman black box.

That's why a component such as the one shown in *Figure 7.7* is called an **oracle**. It's a piece of circuitry that gives us answers even though we don't how it obtained those answers. Is there a CNOT gate inside the box? Nora knows, but we don't.

While you were reading the previous paragraph, Nora was in her lab creating oracles for each of the four simple binary functions. The upcoming subsections tell us what happened during that lab session.

The Zero oracle

If $f(x)$ is always 0, then $f(x) \oplus y = 0 \oplus y = y$. So, we can get the behavior shown in *Figure 7.14* without using any gates (see *Figure 7.16*):

$$|x\rangle \quad \boxed{} \quad |x\rangle$$
$$|y\rangle \quad \phantom{\boxed{XXXX}} \quad |y\rangle = |0 \oplus y\rangle = |f(x) \oplus y\rangle$$

Figure 7.16 – An oracle for the Zero function

The oracle in *Figure 7.16* doesn't do anything. It outputs exactly what it inputs. The matrix representation of circuitry that does nothing is provided here:

$$I = \begin{pmatrix} 1 & 0 & 0 & 0 \\ 0 & 1 & 0 & 0 \\ 0 & 0 & 1 & 0 \\ 0 & 0 & 0 & 1 \end{pmatrix}$$

This is our old friend, the identity matrix.

The One oracle

If $f(x)$ is always 1, we can get the behavior shown in *Figure 7.14* using the circuitry in *Figure 7.17*:

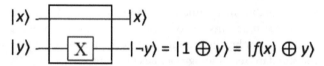

Figure 7.17 – An oracle for the One function

Notice how we use an X gate to turn $|y\rangle$ into $|\neg y\rangle$. To get this circuit's matrix representation, we form a tensor product, like so:

$$X \otimes I = \begin{pmatrix} 0 & 0 & 1 & 0 \\ 0 & 0 & 0 & 1 \\ 1 & 0 & 0 & 0 \\ 0 & 1 & 0 & 0 \end{pmatrix}$$

This matrix is its own inverse, so the circuit is reversible.

The Same_as oracle

If $f(x) = x$, we can get the behavior shown in *Figure 7.14* using the circuitry in *Figure 7.18*:

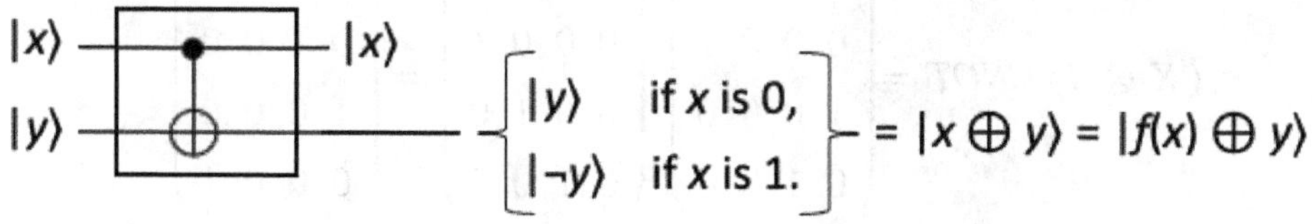

$$|x\rangle \quad \longrightarrow \quad |x\rangle$$
$$|y\rangle \quad \longrightarrow \quad \begin{cases} |y\rangle & \text{if } x \text{ is } 0, \\ |\neg y\rangle & \text{if } x \text{ is } 1. \end{cases} = |x \oplus y\rangle = |f(x) \oplus y\rangle$$

Figure 7.18 – An oracle for the Same_as function

Notice how a CNOT gate implements the XOR operator. When $|x\rangle$ is 0, the bottom qubit's value goes unchanged. When $|x\rangle$ is 1, the bottom qubit's value gets flipped.

A matrix representation of this oracle is provided here:

$$CNOT = \begin{pmatrix} 1 & 0 & 0 & 0 \\ 0 & 0 & 0 & 1 \\ 0 & 0 & 1 & 0 \\ 0 & 1 & 0 & 0 \end{pmatrix}$$

The CNOT matrix is its own inverse, so the *Same_as* oracle is reversible.

> **Important note**
>
> In this chapter, my naming of the binary functions is non-standard. Most authors use the terms *Identity function* and *Identity oracle* instead of *Same_as function* and *Same_as oracle*. But, as you see in the previous paragraphs, the identity matrix implements the *Zero* oracle, not the *Identity* oracle. So, the standard terminology can be confusing.

The Opposite_of oracle

If $f(x) = \neg X$, we can get the behavior shown in *Figure 7.14* using the circuitry in *Figure 7.19*:

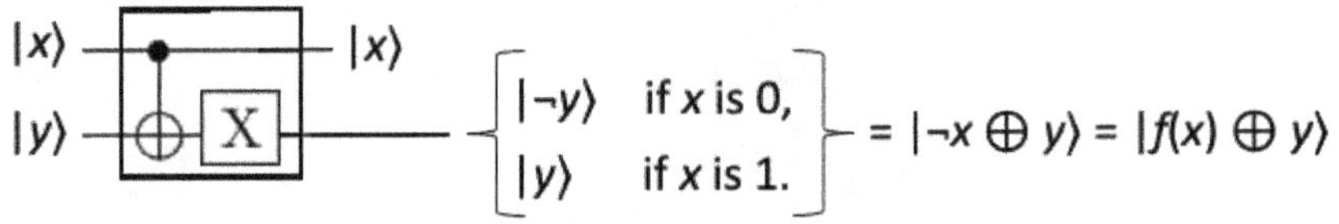

Figure 7.19 – An oracle for the Opposite_of function

This oracle's bottom output is the opposite of the *Same_as* oracle's output. So, to implement this oracle, we add an X gate to the *Same_as* oracle's CNOT gate.

To get this circuit's matrix representation, we do some matrix arithmetic:

$$(X \otimes I)\, CNOT = \begin{pmatrix} 0 & 0 & 1 & 0 \\ 0 & 0 & 0 & 1 \\ 1 & 0 & 0 & 0 \\ 0 & 1 & 0 & 0 \end{pmatrix} \begin{pmatrix} 1 & 0 & 0 & 0 \\ 0 & 0 & 0 & 1 \\ 0 & 0 & 1 & 0 \\ 0 & 1 & 0 & 0 \end{pmatrix} = \begin{pmatrix} 0 & 0 & 1 & 0 \\ 0 & 1 & 0 & 0 \\ 1 & 0 & 0 & 0 \\ 0 & 0 & 0 & 1 \end{pmatrix}$$

As with the other matrices in this section, the *Opposite_of* oracle's matrix is its own inverse. So, the oracle's operation is reversible.

Notice that the oracles for the *Zero* and *One* functions don't have CNOT gates. But the oracles for the two balanced functions, *Same_as* and *Opposite_of*, have CNOT gates. This is the heart and soul of Deutsch's algorithm.

Putting it all together

Let's summarize what we know from the previous sections in this chapter:

- With phase kickback, we can devise a circuit whose topmost output is either $|+\rangle$ or $|-\rangle$.

  If the circuit contains no CNOT gate, the output is $|+\rangle$.

  If the circuit contains a CNOT gate, the output is $|-\rangle$.

- We can distinguish $|+\rangle$ from $|-\rangle$ by applying an H gate and then measuring the qubit.

  If the qubit's state is $|+\rangle$, the measurement we get is 0.

  If the qubit's state is $|-\rangle$, the measurement we get is 1.

- We can implement all four of the simple binary functions with oracles.

  If the oracle has no CNOT gate, it represents a constant function.

  If the oracle has a CNOT gate, it represents a balanced function.

Your friend, Nora, hands you a tiny component and tells you that it implements one of the four simple binary functions. You hook Nora's component up to the circuitry shown in *Figure 7.8* and run the circuit.

If the output is 0, the top qubit's state before the measurement was $|+\rangle$, so Nora's component contains no CNOT gate. Therefore, Nora's component implements a constant function.

If the output is 1, the top qubit's state before the measurement was $|-\rangle$, so Nora's component contains a CNOT gate. Therefore, Nora's component implements a balanced function.

> **Important note**
>
> Using Deutsch's algorithm, you discover whether Nora's component implements a constant or balanced function. If all goes well, you run your circuitry only once. And here's the best part: *Nora's oracle computes the value of f(x) only once.*

At the start of this chapter's *Deutsch's problem* section, I had to unwrap two gifts to find out if they contained the same action figure or different action figures. In the corresponding binary function problem, a classical computer must evaluate the function twice to find out if it's constant or balanced.

But with a quantum computer, you run Nora's oracle only once, and the oracle evaluates the function only once. It's like magic! At its very core, Deutsch's algorithm works because of superposition. When a qubit is in the $|+\rangle$ or $|-\rangle$ state, the qubit has properties belonging to both of the function's possible inputs, – 0 and 1.

Deutsch's algorithm F.A.Q.

Like most people, you may view Deutsch's algorithm with a bit of skepticism. Is this all we can do with a multi-million-dollar quantum computer? Let's consider such questions:

- With Deutsch's algorithm, you evaluate $f(x)$ once instead of twice. Is that time-saving such a big deal?

  The time saving is tiny, but tiny time savings add up when you're doing billions of calculations. Besides, the fact that we can evaluate a function only once and discover something about two possible output values ($f(0)$ and $f(1)$) is amazing. You have to admit that.

- Deutsch's algorithm requires circuitry that's not needed in the classical algorithm. You need a few Hadamard gates and an oracle with two outputs. Does this mean that Deutsch's algorithm is less efficient than the classical algorithm?

  To measure efficiency, computer scientists consider the relationship between the amount of input and the number of times a certain operation is performed. For Deutsch's algorithm, the binary function with two possible inputs (0 or 1) is evaluated only once. That two-to-one ratio is a significant conceptual leap.

- You say that the circuit for Deutsch's algorithm evaluates $f(x)$ once. But the circuit never measures the qubit whose value includes $f(x)$. That $|f(x) \oplus y\rangle$ qubit keeps going until it falls off the circuit's virtual cliff. What's going on here?

  Let's face it. Phase kickback is weird. In the circuit for Deutsch's algorithm, phase kickback takes the payload from the bottom qubit and inserts it into the top qubit. One way or another, Deutsch's algorithm depends on the evaluation of $f(x)$. If we didn't have to evaluate $f(x)$, we wouldn't need four oracles, so we wouldn't need CNOT gates, and the whole algorithm would fall apart.

- Deutsch's algorithm tells me whether a binary function is constant or balanced. The algorithm is efficient, but your examples involving kids' action figures and Nora's mysterious circuit aren't realistic. Why would I ever need to find out whether a function is constant or balanced?

  Please, stop complaining about my examples! Even if there's no practical use for Deutsch's algorithm, the result is a work of beauty. Besides, you never know when abstract concepts will surprise everyone and turn into useful tools. Practitioners weren't interested in prime number theory until it started driving the world's cryptography schemes. No one thought a fourth dimension would be useful until Minkowski used it to describe Einstein's relativity theory. Neural nets weren't very important until they revolutionized machine learning.

- Does Deutsch's algorithm work the same way on simulators and real quantum computers?

  I wish you hadn't asked that question. To run Deutsch's algorithm on a simulator, you need only one shot. But to run the algorithm on a real quantum computer, you need many shots. Today's noisy quantum computers don't always get the right answer. Sometimes, when I run

Deutsch's algorithm with 1,000 shots on a real quantum computer, fewer than 900 of the shots are correct. That's a very poor record for an algorithm that promises one function evaluation instead of two.

But don't fret. Quantum computers are becoming less noisy. And, as circuits for quantum algorithms go, the circuitry for Deutsch's algorithm is fairly simple. It won't be long before quantum computers can perform Deutsch's algorithm in just one shot.

- Can we extend Deutsch's algorithm to functions other than the four simple binary functions?

Yes, we can. For example, the following function of two binary variables is balanced:

$$f(0, 0) = 1$$
$$f(0, 1) = 0$$
$$f(1, 0) = 0$$
$$f(1, 1) = 1$$

The function's oracle and some surrounding circuitry are shown in *Figure 7.20*:

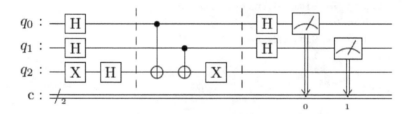

Figure 7.20 – Analyzing a function of two variables

This circuit's output is 01. For a constant function, the output would be 00. For a balanced function, the output would be anything other than 00.

The circuit shown in *Figure 7.20* implements the **Deutsch-Jozsa algorithm**—a generalization of Deutsch's algorithm for functions with any number of inputs.

Enough of the details about Deutsch's algorithm! It's time to see some code.

Coding Deutsch's algorithm

Most of the code in this section uses Qiskit features from previous chapters.

First, we have our `import` declarations:

```
from qiskit import QuantumCircuit, execute
from qiskit_ibm_provider import IBMProvider
from enum import Enum
```

To distinguish between the simple binary functions, we create a Python enumeration:

```
class SimpleBinary(Enum):
    ZERO        = 0
    ONE         = 1
    SAME_AS     = 2
    OPPOSITE_OF = 3
```

This chapter's code makes little use of the `SimpleBinary` enumeration. But having an enumeration feels better than just assigning the numbers 0, 1, 2, and 3 to the four functions.

Our `get_oracle` code adds different gates to a circuit depending on which of the simple binary functions we want to implement:

```
def get_oracle(circ, function):
    # if function == SimpleBinary.ZERO:
        # Do nothing
    if function == SimpleBinary.ONE:
        circ.x(1)
    elif function == SimpleBinary.SAME_AS:
        circ.cnot(0, 1)
    elif function == SimpleBinary.OPPOSITE_OF:
        circ.cnot(0, 1)
        circ.x(1)
    return circ
```

The `get_function` code asks the user to choose between the four binary functions:

```
def get_function():
    print('Which function? (0/1/2/3)')
    print(' 0: ZERO')
    print(' 1: ONE')
    print(' 2: SAME_AS')
    print(' 3: OPPOSITE_OF')
    value = input('> ')
    return SimpleBinary(int(value))
```

The next several statements call `get_function` and `get_oracle`, and surround the oracle with the gates to implement Deutsch's algorithm:

```
circ = QuantumCircuit(2, 1)
function = get_function()

circ.x(1)
circ.h(0)
```

```
circ.h(1)
circ.barrier()
circ = get_oracle(circ, function)
circ.barrier()
circ.h(0)
circ.measure(0, 0)
display(circ.draw('latex'))
```

One run of these statements is shown in *Figure 7.21*:

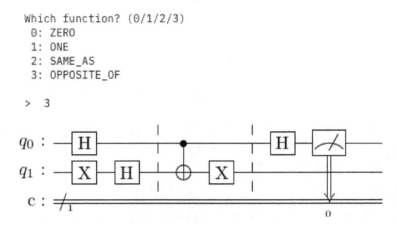

Figure 7.21 – The user selects OPPOSITE_OF

The remaining statements run Deutsch's algorithm on a simulator:

```
provider = IBMProvider()
device = provider.get_backend('ibmq_qasm_simulator')

shots = 1
job = execute(circ, backend=device, shots=shots)
print(job.job_id())

result = job.result()
counts = result.get_counts(circ)

print(function)
print(counts)
number_of_0s = counts.get('0')
number_of_1s = counts.get('1')

if number_of_0s is not None and number_of_0s == shots:
```

```
    print('Constant')
elif number_of_1s is not None and number_of_1s == shots:
    print('Balanced')
else:
    print("Results aren't conclusive")
```

If the user inputs 3 as shown in *Figure 7.21*, the output from these remaining statements is what you see in *Figure 7.22*:

```
SimpleBinary.OPPOSITE_OF
{'1': 1}
Balanced
```

Figure 7.22 – Deutsch's algorithm correctly classifies the Opposite_of function

Notice how cheerfully optimistic we are in this final piece of code. We're running on a simulator, so we expect complete accuracy in the circuit's behavior. For that reason, the code asks for a run with only one shot! If we modified the code by changing `ibmq_qasm_simulator` to the name of a real quantum computer, we'd have to increase the number of shots.

Summary

Deutsch's algorithm doesn't solve problems concerning world hunger or climate change, and it doesn't even print `Hello world`. But the algorithm serves as a proof of concept for the potential of quantum computing. Using Deutsch's algorithm, you can compare two possible input values, 0 and 1, with only one computational step. In the classical world, that would be impossible. It's like seeing inside two boxes by opening only one of them.

In the next chapter, we up the ante by exploring a slightly more complicated quantum algorithm. This new algorithm may not help us develop cold fusion or achieve world peace, but it demonstrates the power of quantum processing over classical computing strategies.

Questions

1. Is this circuit reversible?

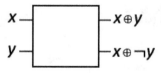

2. In this chapter's *Coding Deutsch's algorithm* section code, change the number of shots from 1 to 100. Then, run the modified code on a real quantum computer. How many of those 100 shots give you the correct answer (constant versus balanced)?

Run the modified code more than once. Observe the variation in the number of correct shots from one run to another.

3. When we run Deutsch's algorithm, the result depends on whether or not there's a CNOT gate inside the oracle. But the result doesn't depend on whether or not there's an X gate in the oracle. Why not?

4. Does the following circuitry implement the *Opposite_of* function? Why, or why not?

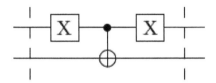

5. Look up the `compose` method belonging to the `QuantumCircuit` class in the Qiskit documentation.

    Modify the `get_oracle` code in this chapter's *Coding Deutsch's algorithm* section so that `get_oracle` returns a separate circuit—a component containing only the gates inside the oracle. After calling `get_oracle`, use the `compose` method to add your oracle to the existing Deutsch's algorithm circuitry.

6. Let $f(x_1, x_2)$ be a binary function with two inputs. How many different ways are there to define $f(x_1, x_2)$? How many of these ways define constant functions? How many define balanced functions? How many are neither constant nor balanced?

7. Let $f(x_1, x_2)$ be a binary function with two inputs. The oracle for such a function looks like this:

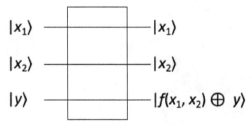

    Describe the circuitry to implement the oracle for the function $f(x_1, x_2) = 1$ (the *One* function).

    Describe the circuitry to implement the oracle for the function $f(x_1, x_2) = x_1$.

    Describe the circuitry to implement the oracle for the function $f(x_1, x_2) = \neg x_2$.

8
Grover's Algorithm

One of my earliest glimpses of quantum computing was in an article about Grover's algorithm. I read the article several times. I understood the mechanics of Grover's algorithm but not the big picture behind it. When you run the algorithm, you start with some qubits, and you change their states. After some amount of fiddling, you measure the system, and (with high probability) the correct answer appears before your eyes! Is this a real algorithm, or is it merely sleight of hand?

Several weeks later, during an hour-long train ride, I scribbled some calculations and convinced myself that the algorithm was destined to work.

The algorithm was originally published in an article entitled *A fast quantum mechanical algorithm for database search*. So, many months later, when I gave a lecture about Grover's algorithm, I told students the algorithm was useful for searching through database records. One of the students asked me how database tables could be mapped to Grover search problems, and I was completely at a loss. I performed some of my usual verbal gymnastics to avoid the question, but the student saw right through my deceit. Much later, I read that Grover had misused the word *database* in the title of his paper.

Despite this grueling embarrassment, Grover's algorithm remains one of my favorite topics in the field of quantum computing. It has more practical uses than Deutsch's algorithm, and it's easier to comprehend than Shor's algorithm. It's a sweet spot for newcomers to the quantum world.

We'll cover the following topics in this chapter:

- How long does it take to find what you need?
- The idea behind Grover's algorithm
- Matrices for Grover's algorithm
- When to use Grover's algorithm
- Gates and circuits for Grover's algorithm
- What does π have to do with Grover's algorithm?

How long does it take to find what you need?

After moving from one town to another, you have 64 unlabeled boxes. One of the boxes contains your coffee pot, and you desperately need a cup of coffee. You line up the boxes in no particular order and open the first box in line. No luck. If your coffee pot were in that box, you'd have seen it right away.

You open the second box, then the third, and then the fourth. Still, no luck. If your worst fear comes true, you won't find the pot until you've opened the 64th box. But chances are, the coffee pot is somewhere in the middle of the line – maybe the 32nd box. That's still not encouraging.

What if you could turn your search into a computer programming problem? Instead of opening boxes, you search an unordered list for the words "coffee pot"? If the list has 64 elements, you may have to check all 64 items. On average, you'll check about half of those items – roughly 32 of them.

But what if you could run your search on a quantum computer? In 1996, a researcher named Lov Grover discovered that you can search among N things in roughly $\sqrt{N}$ steps. No matter where the coffee pot lives among the 64 possible places, you can find the pot in approximately eight steps. What an improvement!

The idea behind Grover's algorithm

The strategy underlying Grover's algorithm is quite clever. Instead of thinking about 64 boxes the way we did in the previous section, let's imagine that you have only four boxes. This set of four boxes is called the **search space**.

You're a quantum computing enthusiast, so you've electronically coded the contents of these boxes and labeled the boxes $|00\rangle$, $|01\rangle$, $|10\rangle$, and $|11\rangle$. Now your search space consists of the four values $|00\rangle$, $|01\rangle$, $|10\rangle$, and $|11\rangle$. In your quantum computing circuit, you represent these values with two qubits, both of which are in the $\begin{pmatrix} \frac{1}{\sqrt{2}} \\ \frac{1}{\sqrt{2}} \end{pmatrix}$ state. When you take the tensor product, you get $\begin{pmatrix} 1/2 \\ 1/2 \\ 1/2 \\ 1/2 \end{pmatrix}$. Remember that

each of the numbers in this vector is an amplitude. The square of each amplitude is the probability of getting a certain outcome when you measure the two qubits. (See *Figure 8.1*.)

When you measure these qubits, the probability of getting . . .

$$\begin{pmatrix} 1/2 \\ 1/2 \\ 1/2 \\ 1/2 \end{pmatrix}$$

$\longrightarrow \quad \ldots |00\rangle$ is $\left(1/2\right)^2 = 1/4$

$\longrightarrow \quad \ldots |01\rangle$ is $\left(1/2\right)^2 = 1/4$

$\longrightarrow \quad \ldots |10\rangle$ is $\left(1/2\right)^2 = 1/4$

$\longrightarrow \quad \ldots |11\rangle$ is $\left(1/2\right)^2 = 1/4$

Figure 8.1 – A state vector's entries correspond to probabilities of measuring values

Figure 8.2 represents the four amplitudes on a horizontal line:

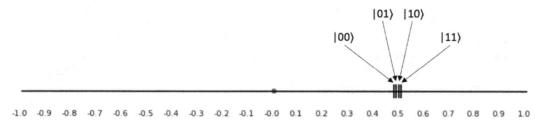

Figure 8.2 – Four amplitudes, each of which is $\dfrac{1}{2}$

In *Figure 8.2*, we can see four vertical lines in slightly different positions. But remember, those four lines represent four amplitudes, each of which is exactly 0.5.

Grover's algorithm has two parts. One of them is the *oracle*, and the other is called the *diffuser*. Let's look at each of them individually.

The oracle

Let's imagine that your coffee pot is in the box labeled $|10\rangle$. We'll call $|10\rangle$ the **target** of the search. What if we have an oracle that knocks the target amplitude to the opposite side of the line? We can see the modified amplitudes in *Figure 8.3*:

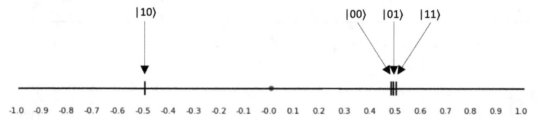

Figure 8.3 – An oracle negates one of the amplitudes

We say that the target is now *marked*. (It's marked by having a minus sign in front of it.) The other three amplitudes remain *unmarked*.

In any application of Grover's algorithm, the oracle always marks the amplitude of the search target by putting a minus sign in front of the target's amplitude.

> **Important note**
>
> When you move from one town to another, you may have two or three coffee pots, and you may have put each of them in a different box. In that scenario, your search space contains three target values. While this can certainly happen, we'll ignore that possibility throughout this chapter. When we perform a search, we assume that the search space contains only one good target value. We could extend Grover's algorithm to perform a search with several target values. But, in this chapter, we'll only deal with a single-target scenario.

The diffuser

Another part of Grover's quantum circuit is called the **diffuser**. *Figure 8.4* illustrates the diffuser's action:

Find the mean (the average) of the four amplitudes:

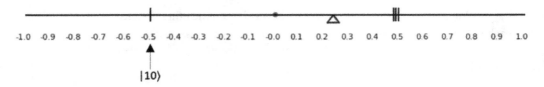

Invert all four of the amplitudes about the mean:

End up with the target value having a probability of 1:

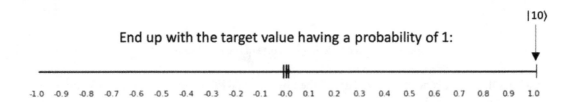

Figure 8.4 – Applying the diffuser

Let's explore the three parts of *Figure 8.4*:

- In the topmost part, a little triangle points to the mean (the average) of the four amplitudes. This mean's value is $\dfrac{\left(\dfrac{1}{2} + \dfrac{1}{2} - \dfrac{1}{2} + \dfrac{1}{2}\right)}{4} = \dfrac{1}{4}$.

- The middle part of *Figure 8.4* illustrates a trick called **inversion about the mean**. The diffuser modifies the qubit's states so that each amplitude's value changes to the mirror image of its earlier value. (Think of the mean as a kind of mathematical mirror.)

- The bottom part of *Figure 8.4* shows the result of the diffuser's action. The value of the target amplitude is 1, and the value of each non-target amplitude is 0. This means that, when you measure the qubits, you'll get $|10\rangle$, which is the label on the box containing your precious coffee pot. Grover's algorithm searches without examining each item.

> **Important note**
>
> I sometimes slip and refer to the diffuser as the *inverter*. Don't follow my lead on that because, if you do, other quantum computing folks might laugh at you.

In this section, we started with two qubits, both in the $|+\rangle$ state. Then, we applied an oracle and the diffuser. Taken together, applying an oracle and the diffuser is called the **Grover iterate**. The word *iterate* suggests repetition. So, in the next section, we'll apply the Grover iterate more than once.

Searching among eight items

In the previous section, we had four items in the search space. From one point of view, each item in the search space was a box full of your possessions. From another point of view, each item in the search space was an amplitude for one of the four possible measurement outcomes – $|00\rangle$, $|01\rangle$, $|10\rangle$, or $|11\rangle$.

Let's repeat the previous section's steps eight items in the search space. To represent eight possibilities, we need three qubits. We say that $n = 3$ and $N = 2^n = 8$. We start with eight amplitudes, each of which is $\frac{1}{\sqrt{8}} \approx 0.35$. (In *Figure 8.5*, imagine that all eight lines are sitting at the same point.)

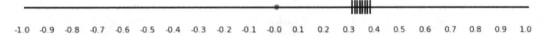

Figure 8.5 – Start with eight amplitudes

One of these amplitudes represents the target. Apply the oracle to move the target to the negative side . (See *Figure 8.6*.)

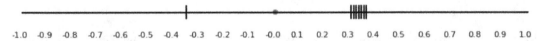

Figure 8.6 – The target is now marked

The mean of the amplitudes is approximately 0.265. We invert all amplitudes about this mean value. (See *Figure 8.7*.)

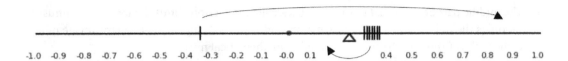

Figure 8.7 – Invert about the mean of the amplitudes

This inversion sends the target amplitude to approximately 0.88, and each non-target amplitude to roughly 0.18. (See *Figure 8.8.*)

Figure 8.8 – The amplitudes after one application of the Grover iterate

Have you found your coffee pot yet? No, not really. The box containing the coffee pot has an amplitude of around 0.88, and the square of that amplitude is roughly 0.78. When you measure this system's qubits, the chance of finding the correct three-bit combination is less than 80%. What can we do about that?

Here's what we can do: we can apply the Grover iterate (the oracle and the diffuser) a second time. Let's give it a go.

The oracle puts a minus sign in front of the target amplitude . (See *Figure 8.9.*)

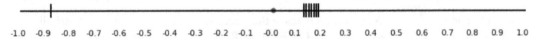

Figure 8.9 – After the second application of the oracle

The mean is approximately 0.044, and the diffuser inverts all amplitudes about this mean. (See *Figure 8.10.*)

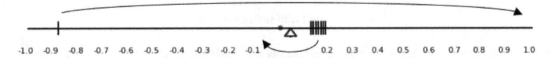

Figure 8.10 – Inversion about the mean

The result after a second application of the Grover iterate is that the non-target amplitudes are approximately –0.09, and the target amplitude is roughly 0.97. (See *Figure 8.11.*)

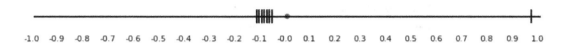

Figure 8.11 – After two applications of the Grover iterate

This translates into a probability of about 0.95 of seeing the target state when you measure all three qubits. That's pretty good for the effort of repeating the Grover iterate only two times.

Searching among any number of items

Table 8.1 contains some useful numbers:

Number of items in the search space:	Number of times you should apply the Grover iterate:
4	1
8	2
. . .	. . .
1,000,000	784
. . .	. . .
N	$\frac{\pi}{4}\sqrt{N}$

Table 8.1 – The optimal number of Grover iterate applications

Let's say you're searching through 1,000,000 items. On average, with classical computing, you have to examine 500,000 items. That's 499,216 steps more than the 784 steps in *Table 8.1*. I've never had 1,000,000 boxes to move. But if I did, I'd want Grover's algorithm by my side.

What happens if we become overzealous and apply the Grover iterate too many times? Let's start where we left off in *Figure 8.4*. We had two qubits with their four amplitudes. After one application of the Grover iterate, the target state's amplitude had a probability of 1. *Figure 8.12* shows what happens when we apply the Grover iterate a second time:

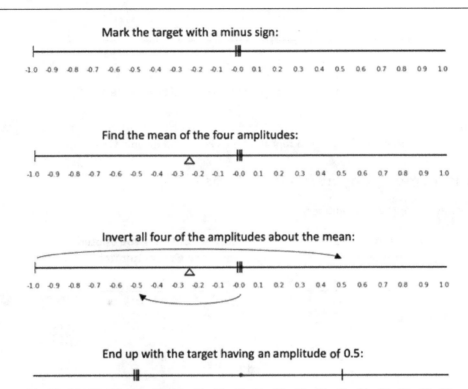

Figure 8.12 – Applying a second iterate when N = 4

After a second application of Grover's iterate, the probability of getting the target when you do a measurement is only $(0.5)^2 = 0.25$. What a disappointment! When Grover told us to apply his iterate $\frac{\pi}{4}\sqrt{N}$ times, he meant it. If you apply the iterate too few or too many times, you mess up the algorithm.

In the next several sections, you'll learn why the number of steps has anything to do with $\sqrt{N}$. Near the end of this chapter, you get a glimpse at the role of $\frac{\pi}{4}$ in the running time calculations.

The optimal number of Grover iterate applications

Figure 8.13 shows the early stage of a Grover search. Here, we've already put all qubits in superposition and applied the oracle one time. We used a solid rectangle to suggest that the number of unmarked amplitudes is very large. But remember, all the unmarked amplitudes have the same value:

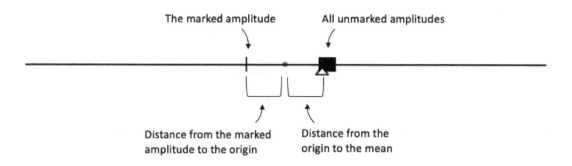

Figure 8.13 – After the first application of an oracle

In *Figure 8.13*, we make the mean look as if it's almost the same as each unmarked amplitude. There's only one marked amplitude and very many unmarked amplitudes. So, the mean stays close to the unmarked amplitudes. For example, imagine that we have 999,999 unmarked amplitudes and one marked amplitude. In this case, each of the unmarked amplitudes is $\frac{1}{\sqrt{1000000}}$, and the marked amplitude is $-\frac{1}{\sqrt{1000000}}$. The mean is $\frac{0.999998}{\sqrt{1000000}}$, which is absurdly close to $\frac{1}{\sqrt{1000000}}$. So, in *Figure 8.13*, we picture the target amplitude as being two distances to the left of the mean.

Figure 8.14 shows the effect of inversion about the mean. After inversion, the target amplitude is approximately two distances to the right of the mean:

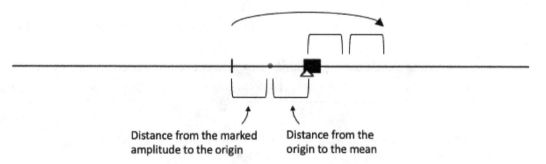

Figure 8.14 – One inversion about the mean

Being two distances to the right of the mean is the same as being three distances from the origin . (See *Figure 8.15*.)

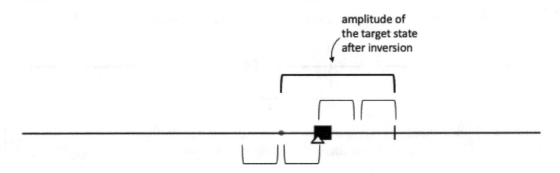

Figure 8.15 – After one application of the Grover iterate

If our search space's size is 1,000,000, we have 1,000,000 amplitudes, and the target amplitude is now very close to $3 \cdot \frac{1}{\sqrt{1000000}}$. If we had N amplitudes, the target amplitude would have been inverted from $-\frac{1}{\sqrt{N}}$ to $\frac{3}{\sqrt{N}}$. No matter how large our search space is, we've applied the Grover iterate once and multiplied the original amplitude by 3. We've done what's called **amplitude amplification**.

> **Important note**
>
> Every time we perform inversion about the mean, the amplitudes of the unmarked states move to the left. The figures in this section don't account for that fact because, for large search spaces, the amount of movement is close to zero.

What happens when we apply the Grover iterate a second time? When we apply the oracle, we go from *Figure 8.15* to *8.16*:

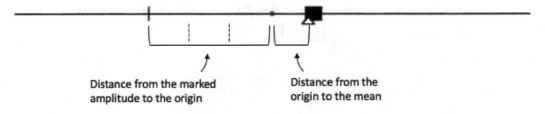

Figure 8.16 – Applying the oracle a second time

In *Figure 8.16*, two dashed lines remind us that the distance from the marked amplitude to the origin is approximately three times the distance from the origin to the mean. In *Figure 8.17*, we apply the diffuser. As always, the diffuser inverts all amplitudes about the mean:

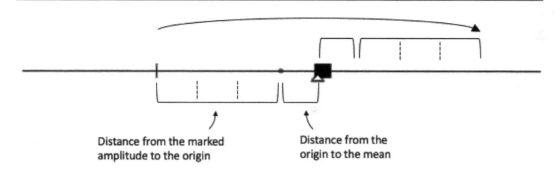

Distance from the marked
amplitude to the origin

Distance from the
origin to the mean

Figure 8.17 – Applying the diffuser a second time

Figure 8.18 shows the target amplitude after a second application of the Grover iterate. Notice how the target amplitude is composed of 5 roughly equal distances. What started as $\frac{1}{\sqrt{N}}$ is now approximately $\frac{5}{\sqrt{N}}$. (You can say that we've just amplified the amplitude amplification!)

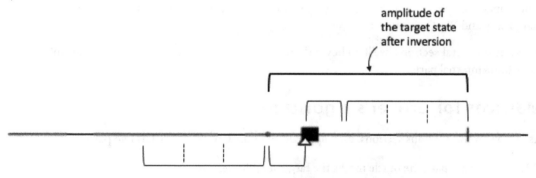

amplitude of
the target state
after inversion

Figure 8.18 – After two applications of the Grover iterate

And so it goes. After one iteration, the target state's amplitude is $\frac{3}{\sqrt{N}}$. After two iterations, the amplitude is $\frac{5}{\sqrt{N}}$. After three iterations, it would be $\frac{7}{\sqrt{N}}$.

In general, after k iterations, the target state's amplitude is $\frac{2k+1}{\sqrt{N}}$.

Remember that we want a nearly 100% probability of seeing the target state when we finally measure these qubits. So, when is the target state's amplitude, $\frac{2k+1}{\sqrt{N}}$, close to 1? Here's the math:

$$\frac{2k+1}{\sqrt{N}} = 1$$

$$2k + 1 = \sqrt{N}$$

$$2k \approx \sqrt{N}$$

$$k \approx \frac{\sqrt{N}}{2}$$

Notice how these calculations don't end with an equal sign. This chapter contains at least a dozen occurrences of the word "approximate." The correct number of Grover iterate applications isn't exactly $\frac{\sqrt{N}}{2}$. Many authors say that the correct number has something to do with $\sqrt{N}$ and leave it at that. A phrase such as "something to do with $\sqrt{N}$" is appropriate because, with a search space of size N, a classical algorithm might have to perform N steps. And N is much larger than $\sqrt{N}$. (If you want to know how to make the phrase "much larger than $\sqrt{N}$" more precise, read some material on computational complexity and Big-O notation.)

In the next several sections, we'll go beyond our overview of Grover's algorithm and examine the algorithm's internal parts.

Matrices for Grover's algorithm

As we saw in the previous sections, each application of the Grover iterate has two parts:

1. In the first part, the oracle marks the target amplitude.

2. In the second part, the diffuser inverts all amplitudes about the mean.

Each part is a collection of quantum gates, and those gates apply a matrix to the circuit's qubits. In this section, we'll describe the matrix representations of the oracle and the diffuser.

A matrix for the oracle

Let's assume that we have only two qubits. After these qubits go through Hadamard gates, we have a matrix representation of $\frac{1}{2} \begin{pmatrix} 1 \\ 1 \\ 1 \\ 1 \end{pmatrix}$. If we want to mark the $|10\rangle$ amplitude, we must do the following:

$$\begin{pmatrix} 1 & 0 & 0 & 0 \\ 0 & 1 & 0 & 0 \\ 0 & 0 & -1 & 0 \\ 0 & 0 & 0 & 1 \end{pmatrix} \frac{1}{2} \begin{pmatrix} 1 \\ 1 \\ 1 \\ 1 \end{pmatrix} = \frac{1}{2} \begin{pmatrix} 1 \\ 1 \\ -1 \\ 1 \end{pmatrix}$$

In general, the oracle's matrix is the identity matrix with one of the 1s along the diagonal changed to a -1. Simple as this is, it's also somewhat disconcerting. To know which diagonal element becomes -1, you have to know which item is the target. And if you already know which item is the target, why bother searching for the target with Grover's algorithm? We'll tackle that question in the *When to use Grover's algorithm* section.

A matrix for the diffuser

Let's start with two qubits. Here, we have four amplitudes in a vector, $\begin{pmatrix} a \\ b \\ c \\ d \end{pmatrix}$. *Figure 8.19* illustrates the inversion of one of those amplitudes about the mean:

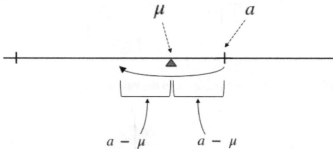

Figure 8.19 – Inverting amplitude, a, about the mean, μ

The mean of the four amplitudes, a, b, c, and d, is as follows:

$$\mu = \frac{a + b + c + d}{4}$$

To invert a about the mean, we must subtract the difference, $a - \mu$, from a twice:

$$a - 2(a - \mu) = a - 2a + 2\mu$$
$$= -a + 2\mu$$
$$= -\frac{2a}{2} + \frac{a + b + c + d}{2}$$
$$= \frac{-a + b + c + d}{2}$$
$$= \frac{1}{2}(-a + b + c + d)$$

Now, consider the following matrix:

$$D = \frac{1}{2}\begin{pmatrix} -1 & 1 & 1 & 1 \\ 1 & -1 & 1 & 1 \\ 1 & 1 & -1 & 1 \\ 1 & 1 & 1 & -1 \end{pmatrix}$$

Here's what happens when we apply this matrix, D, to the pair of qubits:

$$\frac{1}{2}\begin{pmatrix} -1 & 1 & 1 & 1 \\ 1 & -1 & 1 & 1 \\ 1 & 1 & -1 & 1 \\ 1 & 1 & 1 & -1 \end{pmatrix}\begin{pmatrix} a \\ b \\ c \\ d \end{pmatrix} = \frac{1}{2}\begin{pmatrix} -a + b + c + d \\ a - b + c + d \\ a + b - c + d \\ a + b + c - d \end{pmatrix}$$

In the resulting vector, the topmost entry is $\frac{1}{2}(-a + b + c + d)$, and that's what you get when you invert amplitude a about the mean. The same is true for amplitudes b, c, and d in the other rows of the vector. So, the matrix, D, represents the diffuser in a circuit with four amplitudes. We can rewrite the D matrix as follows:

$$\frac{2}{4} \begin{pmatrix} 1 - \dfrac{4}{2} & 1 & 1 & 1 \\[2.2em] 1 & 1 - \dfrac{4}{2} & 1 & 1 \\[2.2em] 1 & 1 & 1 - \dfrac{4}{2} & 1 \\[2.2em] 1 & 1 & 1 & 1 - \dfrac{4}{2} \end{pmatrix}$$

When we do, we get a specific version of the more general N- amplitude diffusion matrix:

$$\frac{2}{N} \begin{pmatrix} 1 - \dfrac{N}{2} & 1 & \cdots & 1 \\[2.2em] 1 & 1 - \dfrac{N}{2} & \cdots & 1 \\[1.5em] \vdots & \vdots & \ddots & 1 \\[1.5em] 1 & 1 & \cdots & 1 - \dfrac{N}{2} \end{pmatrix}$$

With n qubits, this matrix inverts $N = 2^n$ amplitudes about the mean.

It's helpful to think of the diffusion matrix as a tensor product of some more familiar matrices. To do so, start with the identity matrix and change the *upper left* entry from 1 to -1:

$$\begin{pmatrix} -1 & 0 & 0 & \cdots & 0 \\ 0 & 1 & 0 & \cdots & 0 \\ 0 & 0 & 1 & \cdots & 0 \\ \vdots & \vdots & \vdots & \ddots & \vdots \\ 0 & 0 & 0 & \cdots & 1 \end{pmatrix}$$

This matrix alone has the effect of flipping the $|00 \cdots 0\rangle$ amplitude. When you surround this matrix with Hadamard gates, you get a diffuser matrix. Here's how it works for a two-qubit circuit:

$$(H \otimes H) \begin{pmatrix} -1 & 0 & 0 & 0 \\ 0 & 1 & 0 & 0 \\ 0 & 0 & 1 & 0 \\ 0 & 0 & 0 & 1 \end{pmatrix} (H \otimes H)$$

$$= \frac{1}{4} \begin{pmatrix} 1 & 1 & 1 & 1 \\ 1 & -1 & 1 & -1 \\ 1 & 1 & -1 & -1 \\ 1 & -1 & -1 & 1 \end{pmatrix} \begin{pmatrix} -1 & 0 & 0 & 0 \\ 0 & 1 & 0 & 0 \\ 0 & 0 & 1 & 0 \\ 0 & 0 & 0 & 1 \end{pmatrix} \begin{pmatrix} 1 & 1 & 1 & 1 \\ 1 & -1 & 1 & -1 \\ 1 & 1 & -1 & -1 \\ 1 & -1 & -1 & 1 \end{pmatrix}$$

$$= \frac{1}{4} \begin{pmatrix} 1 & 1 & 1 & 1 \\ 1 & -1 & 1 & -1 \\ 1 & 1 & -1 & -1 \\ 1 & -1 & -1 & 1 \end{pmatrix} \begin{pmatrix} -1 & -1 & -1 & -1 \\ 1 & -1 & 1 & -1 \\ 1 & 1 & -1 & -1 \\ 1 & -1 & -1 & 1 \end{pmatrix}$$

$$= \frac{1}{4} \begin{pmatrix} 2 & -2 & -2 & -2 \\ -2 & 2 & -2 & -2 \\ -2 & -2 & 2 & -2 \\ -2 & -2 & -2 & 2 \end{pmatrix}$$

$$= \frac{1}{2} \begin{pmatrix} 1 & -1 & -1 & -1 \\ -1 & 1 & -1 & -1 \\ -1 & -1 & 1 & -1 \\ -1 & -1 & -1 & 1 \end{pmatrix} = \text{the 4-amplitude diffusion matrix}$$

The next section will turn these matrices into Qiskit code.

Coding Grover's algorithm with matrices

This chapter contains three different ways to code Grover's algorithm. The first way uses matrices from the previous section. As usual, we begin with the imports:

```
from qiskit import QuantumCircuit, Aer, execute
from qiskit.visualization import plot_histogram
```

In *Chapter 1*, we used NumPy to do matrix arithmetic in Python. In this chapter, we don't need Python for matrix arithmetic. We need matrices to define custom quantum gates, but our matrices can be ordinary Python lists. Here's some code:

```
oracle_matrix = [
    [1, 0,  0, 0, 0, 0, 0, 0],
    [0, 1,  0, 0, 0, 0, 0, 0],
    [0, 0,  1, 0, 0, 0, 0, 0],
    [0, 0,  0, 1, 0, 0, 0, 0],
    [0, 0,  0, 0, -1, 0, 0, 0],
    [0, 0,  0, 0, 0, 1, 0, 0],
    [0, 0,  0, 0, 0, 0, 1, 0],
    [0, 0,  0, 0, 0, 0, 0, 1]
]
oracle = QuantumCircuit(3)
oracle.unitary(oracle_matrix, qubits=[0, 1, 2], label='oracle')
oracle.barrier()
display(oracle.draw('latex'))
```

The matrix in this code defines an oracle to mark the $|100\rangle$ amplitude. We turn this matrix into a circuit with Qiskit's `unitary` method. The `unitary` method's three parameters include the matrix (of course), a `label` parameter to display in drawings of the circuit, and a list of `qubits`. In this code, [0, 1, 2] tells Qiskit to apply the matrix to qubits 0, 1, and 2, in that order (the most natural order). As a result, the oracle's circuit drawing looks like the one in *Figure 8.20*:

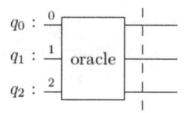

Figure 8.20 – Displaying the newly created gate

A call to the `unitary` method with the `qubits=[2, 0, 1]` parameter would have spawned a slightly different drawing and would have applied the matrix's minus sign to $|101\rangle$ instead of $|100\rangle$. (Check this for yourself!)

The following piece of code creates the algorithm's diffuser:

```
flip_matrix = [
    [-1, 0, 0, 0, 0, 0, 0, 0],
    [ 0, 1, 0, 0, 0, 0, 0, 0],
    [ 0, 0, 1, 0, 0, 0, 0, 0],
    [ 0, 0, 0, 1, 0, 0, 0, 0],
    [ 0, 0, 0, 0, 1, 0, 0, 0],
    [ 0, 0, 0, 0, 0, 1, 0, 0],
    [ 0, 0, 0, 0, 0, 0, 1, 0],
```

```
    [ 0, 0, 0, 0, 0, 0, 0, 1]
]
flip = QuantumCircuit(3)
flip.unitary(flip_matrix, qubits=[0, 1, 2], label='flip')
h3 = QuantumCircuit(3)
h3.h([0, 1, 2])

diffuser = h3.compose(flip).compose(h3)
diffuser.barrier()
display(diffuser.draw('latex'))
```

Like the code to create an oracle, this diffuser code uses `unitary` to make a circuit from a matrix. And here's something new: you can put a list of qubits inside a method call to add more than one gate to a circuit. That's why we can abbreviate three separate method calls (`h(0)`, `h(1)`, and `h(2)`) with one `h([0, 1, 2])` call.

In Qiskit, the `compose` method pastes one circuit to the end of another circuit. So, in this example, we paste `flip` to a circuit with three Hadamard gates, and then paste three more Hadamard gates after `flip`. The combined circuit (in *Figure 8.21*) forms the diffuser for any three-qubit Grover algorithm:

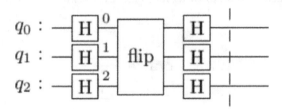

Figure 8.21 – The diffuser

The two methods, `unitary` and `compose`, have a lot in common. Both methods add things to existing circuits. In addition, both methods have `qubits` parameters that specify the order in which parts should be combined. With the `unitary` method, the `qubits` parameter is required.

The following part of the code combines the oracle and the diffuser into one circuit:

```
grover_iterate = oracle.compose(diffuser)
```

When I started writing this code, I didn't bother defining `grover_iterate`. But, in Grover's algorithm, you never apply the oracle and diffuser separately. So, why bother keeping track of both the oracle and the diffuser? Instead, turn the two of them into one lump of gates and add this lump as many times as you need.

We're finally ready to define the complete circuit for Grover's algorithm on three qubits. Here's the code:

```
circ = QuantumCircuit(4, 3) # We use the fourth qubit
                    #    later in this chapter.
circ.h([0, 1, 2])
circ.barrier()
circ = circ.compose(grover_iterate).compose(grover_iterate)
circ.measure([0, 1, 2], [0, 1, 2])
display(circ.draw('latex'))
```

The circuit that this code creates is shown in *Figure 8.22*:

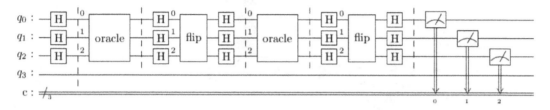

Figure 8.22 – Grover's algorithm for n = 3, N = 2³ = 8

The code to run this circuit contains no surprises:

```
device = Aer.get_backend('qasm_simulator')
job = execute(circ,backend = device,shots = 1000)
print(job.job_id())

result = job.result()
counts = result.get_counts(circ)

print(counts)
display(plot_histogram(counts))
```

When I ran the circuit in *Figure 8.22*, I got the output shown in *Figure 8.23*:

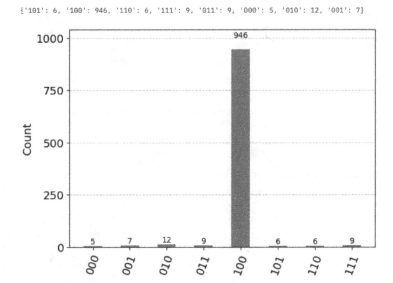

{'101': 6, '100': 946, '110': 6, '111': 9, '011': 9, '000': 5, '010': 12, '001': 7}

Figure 8.23 – We found |100⟩

As Grover's algorithm predicts, the measurement outcome, |100⟩, has a very high probability. Inverting about the mean has lived up to its reputation!

It's time to look at some uses for Grover's algorithm.

When to use Grover's algorithm

Before signing a contract to use Grover's algorithm, you must read the fine print. To run Grover's algorithm, you need an oracle that marks your search's target amplitude. Think about the analogy at the start of this chapter. You have 64 boxes, and you wonder which box contains your coffee pot. Along comes an omniscient oracle who knows which box contains the pot. The oracle puts a mark on that box for everyone to see.

The oracle "knows" which item is the target and marks that item. But, in quantum computing, someone or something has to write the oracle's code. Why can't we just ask that code-writing agent which item it marked? Why bother doing inversion about the mean? Why trouble yourself by applying the Grover iterate?

For Grover's algorithm to be useful, a search problem must have certain characteristics. Among them is the requirement that you can code the oracle without knowing the problem's target value. Here's some code whose role is like that of a quantum oracle:

```
x = float(input())
if x**5 - 2*(x**4) + 4*(x**3) - 8*(x**2) + 3*x - 6 == 0:
```

```
      print(1)
  else:
      print(-1)
```

This code prompts you for a number and displays 1 or -1. It displays -1 only when the number you enter is 2. (Of course, I'm ignoring numbers such as 2.0000000000000001, which the computer mistakes for 2.0.) This code displays -1 only for solutions to the equation:

$$x^5 - 2x^4 + 4x^3 - 8x^2 + 3x - 6 = 0$$

Let's say this was a quadratic equation, like so:

$$x^2 + 4x + 4 = 0$$

Here, you could apply the quadratic formula to find solutions. But it's not a quadratic equation. It's a fifth-degree equation and, in the 1800s, mathematicians proved that there's no general formula for solving fifth-degree equations. The Python code containing this equation can easily recognize that 2 is a solution. But there's no easy formula to find a solution if you don't already have one. Code that *recognizes* a correct input doesn't necessarily come from code that *finds* that correct input from scratch.

The next few sections describe some problems whose oracles can be written without us knowing their target values.

Encrypting passwords

My computer's /etc/passwd file contains the following line:

```
root:*:0:0:System Administrator:/var/root:/bin/sh
```

On this line, the asterisk stands for my system's root password. The operating system doesn't show me the actual password because the system doesn't store that password. Instead, the system stores an encrypted version of that password. If my password is swordfishdavidcoconut, the computer encrypts this and stores something like xyJXTYIlXthPFBiX7kpdctk%.

There are many different ways to encrypt passwords. The method used by my operating system is one-way encryption, making it easy to encrypt a password and nearly impossible to decrypt the encrypted text:

- When the system requests my password, I type swordfishdavidcoconut. The system encrypts what I type and compares the encrypted value with xyJXTYIlXthPFBiX7kpdctk%. If there's a match, the system lets me in.

- If I forget my password, it's nearly impossible to get it back from the system. The best approach would be to do what hackers do: they make one guess after another until they stumble on the correct input. Assuming that the password contains only lowercase Roman letters, it could take 26^{21} guesses to guess `swordfishdavidcoconut`. At one guess per nanosecond, that would take 10^{13} years.

When I try to log on, the code that compares my input with the encrypted text is a kind of oracle. It doesn't put a minus sign in front of my correct input, but it does essentially the same thing. It does this without being able to guess what the correct input would be. That makes encrypting passwords a candidate for a solution using Grover's algorithm. Grover's algorithm uses superposition to evaluate all 26^{21} possible guesses in parallel. With Grover's algorithm, we can reduce the time to guess a 21-letter password from 10^{13} years to 10^6 years.

(What's that voice in your head saying? Isn't 10^6 years good enough for you? Stop being so cynical! Quantum computers are still in their infancy. Today's best Intel chip is 50,000 times faster than the original ENIAC computer. Who knows what the future will bring?)

Let's look at another problem that lends itself to a solution using Grover's algorithm.

Finding better approximations

Every year, the Affordable Motor Company's executives must decide on the number of cars of each type to manufacture. The company's analysts write a series of equations to describe the anticipated profit stemming from each combination of cars. They write computer code to find the optimal solution for the equations. But, if they waited for the code to finish running, it would be time to build the following year's models.

Instead of looking for an optimal solution with the most profit, the analysts write code to find approximate solutions. The code finds a solution of any kind (a candidate solution), then a slightly better solution (another candidate), then an even better solution, and so on. At some point, they stop running the code and deliver the most recent candidate solution to the company's CEO. (The CEO rejects this solution because of an unexplainable gut feeling, but that's another matter.)

In this scenario, an oracle stores information about the best candidate that it's found so far. Then, the oracle considers many alternative candidates. To decide whether to mark an alternative candidate, the oracle doesn't try to solve the analysts' equations. Instead, the oracle plugs an alternative candidate's values into the analysts' equations. Plugging into these equations is a relatively easy calculation. The oracle marks an alternative candidate if the alternative's anticipated profit is higher than the best-so-far profit.

Grover's algorithm can find the marked solutions much faster than a classical-computing algorithm.

Satisfying Boolean expressions

Consider the well-known **satisfiability problem** (**SAT**). With this problem, someone hands you a Boolean expression and asks you whether the expression can *possibly* be true. A Boolean expression that can possibly be true is said to be **satisfiable**. Here's an example:

Alice, Bob, and Eve share a large pizza. Alice wants anchovies, pineapple, or both. Bob wants pineapple, sausage, or both. Eve hates anchovies. She doesn't mind sausage or pineapple, but won't eat them together on a pizza.

If you abbreviate each ingredient with its first letter, you get the following expression:

```
(a or p) and (p or s) and (not a) and (not p or not s)
```

To satisfy this expression, let a be false, p be true, and s be false. Get pineapple on the pizza – no anchovies and no sausage.

Here's another expression:

```
x and not x
```

There's no way to make this expression true. This expression isn't satisfiable.

The SAT problem belongs to a category of problems that are **NP-complete**. There's a rigorous definition of NP-completeness, but the formalities aren't important here. Suffice it to say that the SAT problem shares three characteristics with all other NP-complete problems. Here are the first two characteristics:

- If someone proposes a combination of Boolean values that satisfy a particular expression, there's an efficient way to check whether that combination satisfies the expression
- As far as we know, if you don't already have a combination of Boolean values that satisfy a particular expression, there's no efficient way to find such a combination

The assertions in these first two bullets suggest that the SAT problem is ripe to be solved by Grover's algorithm. Here's the third characteristic of the SAT problem:

- Many difficult problems can be re-worded as SAT problems

You name the problem: the traveling salesperson problem, the map coloring problem, some job scheduling problems, the vehicle routing problem, some problems involving the game Tetris, and... yes... even Alice, Bob, and Eve's pizza ordering problem. All of these problems can be expressed as strings of Boolean variables.

The speedup that we get with Grover's algorithm applies to a wide variety of problems. So, let's write some code!

Coding Grover's algorithm with high-level functions

Some third-party code libraries can do most of Grover's work for you. A library named **tweedledum** creates an oracle from a Boolean expression, and the **qiskit_algorithms** library builds Grover iterates from an oracle.

You can install `tweedledum` by running the following instruction in your Jupyter notebook:

```
pip install tweedledum
```

When you run this instruction, the output may tell you to restart the notebook's kernel. After a restart, `tweedledum` is available in your notebook. But that availability doesn't last forever. When you start to work the next day, you'll probably have to install `tweedledum` again.

In another cell of your Jupyter notebook, run the following instruction:

```
pip install qiskit_algorithms
```

Now you're ready to use these libraries.

Your next step is to run the imports:

```
from qiskit.circuit.library.phase_oracle import PhaseOracle
from qiskit_algorithms import AmplificationProblem, Grover
from qiskit.tools.visualization import plot_histogram
```

In this example, we'll code the following pizza topping constraints:

> *Alice wants sausage, no anchovies, and pineapple. Bob wants mushrooms or anchovies (or both).*

In the expression that we feed to `PhaseOracle`, we represent `not` with a tilde (~), and with an ampersand (&), and `or` with a pipe (|). If we need to use an exclusive or, we represent it with a caret (^). Given these abbreviations, we write the code for Alice's and Bob's pizza preferences:

```
expression = ('(sausage & ~anchovies & pineapple)' \
              ' & (mushrooms | anchovies)')
print(expression)

oracle = PhaseOracle(expression)
problem = AmplificationProblem(oracle)
grover = Grover(iterations=2)
circ = grover.construct_circuit(problem)
circ.measure_all()
display(circ.draw('latex'))
```

From `expression`, we get `oracle`. From `oracle`, we get `problem`. From `problem` and the `Grover` instance, we get a circuit. The expression and the circuit are shown in *Figure 8.24*:

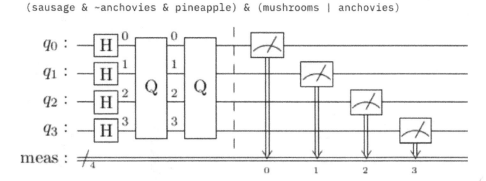

Figure 8.24 – A ready-made Grover circuit

Each qubit in the circuit stands for an ingredient in the Boolean expression. To find the role of each qubit, list the ingredients in the expression in order of appearance. When you do this, you find that q_0 stands for sausage, q_1 for anchovies, q_2 for pineapple, and q_3 for mushrooms.

Running the circuit is nothing new. I've copied the code here:

```
from qiskit import Aer, execute
from qiskit.visualization import import plot_histogram

device = Aer.get_backend('qasm_simulator')

job = execute(circ,backend = device, shots = 1000)
print(job.job_id())

result = job.result()
counts = result.get_counts(circ)

print(counts)
display(plot_histogram(counts))
```

After running this code, I got the output shown in *Figure 8.25*:

{'1111': 5, '1010': 6, '1000': 5, '0111': 8, '0000': 4, '0011': 6, '1011': 6, '1001': 8, '1110': 6, '1101': 914, '0010': 8, '0001': 4, '0110': 6, '0100': 5, '0101': 3, '1100': 6}

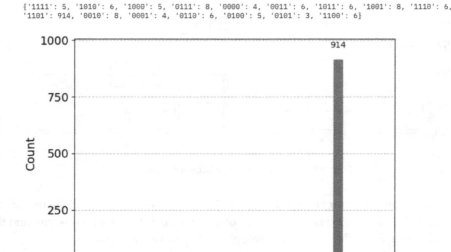

Figure 8.25 – At last, Alice and Bob know what to put on their pizza!

In *Figure 8.25*, the winning amplitude, $|1101\rangle$, means "mushrooms, pineapple, no anchovies (thank goodness!), and sausage."

Gates and circuits for Grover's algorithm

Qiskit's `PhaseOracle` and `Grover` classes provide turnkey solutions to complicated problems. Turnkey solutions can get results quickly and painlessly, but solutions of this kind have limitations. Turnkey solutions seldom apply to problems that have unusual constraints, and unusual constraints pop up often in real life. What's more, when you apply a turnkey solution, you gain little or no insight about the way you solved the problem.

So, in this section, we will drill down into the circuitry to implement Grover's algorithm. We will focus on satisfiability because it applies to so many kinds of problems.

Gates for the oracle

In *Chapter 7*, we introduced a circuit to demonstrate **phase kickback**. I've copied a drawing of the circuit here:

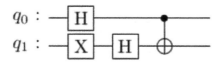

Figure 8.26 – The target kicks the change back to its source

Phase kickback isn't limited to gates with two qubits. *Figure 8.27* contains a three-qubit Toffoli gate that experiences phase kickback:

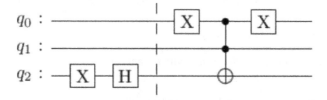

Figure 8.27 – Phase kickback on a larger scale

The circuit in *Figure 8.27* kicks a change back to the two source qubits – q_1 and q_0. The result is a sign flip for the $|10\rangle$ amplitude of those two source qubits. It's the $|10\rangle$ amplitude because, in *Figure 8.27*, q_0 has a pair of X gates and q_1 doesn't. If both qubits had a pair of X gates, the circuit would flip the $|00\rangle$ amplitude. If only q_1 had a pair of X gates, the circuit would flip the $|01\rangle$ amplitude.

As we'll see later in this chapter, Qiskit has a method that surrounds a source qubit with a pair of X gates. For now, don't worry about the method. Just look at *Figure 8.28*:

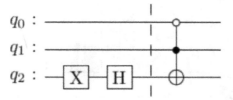

Figure 8.28 – A diagram that's equivalent to the one in Figure 8.27

Qiskit also gives us a way to define control gates with more than two source qubits. For example, *Figure 8.29* shows a circuit that flips the $|101\rangle$ amplitude for three qubits:

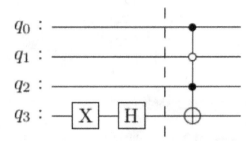

Figure 8.29 – A circuit to recognize |101⟩

Now, consider this important, real-life problem:

> *Alice wants sausage, no anchovies, and some pineapple. Bob wants mushrooms and (like all sensible people) despises anchovies.*

Here's a Boolean expression to represent Alice's and Bob's desires:

```
(s and not a and p) and (m and not a)
```

For a circuit that *almost* realizes this expression, see *Figure 8.30*:

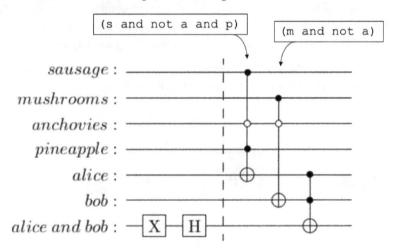

Figure 8.30 – Almost an oracle (but not quite)

In *Figure 8.30*, the **alice** qubit monitors the truth or falsehood of (s and not a and p), the **bob** qubit monitors the truth or falsehood of (m and not a), and the **alice and bob** qubit checks to see whether both Alice and Bob get their wishes.

The circuit in *Figure 8.30* would be an oracle for our Boolean pizza expression except for one tiny problem – a problem whose focus is the **alice** and **bob** qubits. These two qubits are scratch qubits.

They store temporary values. Whatever values they have won't be used later in the computation. But in subsequent applications of the Grover iterate, we'll still need scratch qubits. Instead of wasting the precious **alice** and **bob** qubits, we want to reuse them in subsequent steps. So, we must make sure that **alice** and **bob** haven't been tainted in some way by the circuit in *Figure 8.30*. And yet, **alice** and **bob** have become tainted. Here's why.

We start Grover's algorithm by applying H gates to the four toppings qubits. In the resulting superposition of states, one of the combinations that's represented is *sausage, mushrooms*, no *anchovies*, and *pineapple*. This combination makes both Alice and Bob happy. So, for that part of the superposition, the **alice** and **bob** qubits are both $|1\rangle$. (See *Figure 8.31.*)

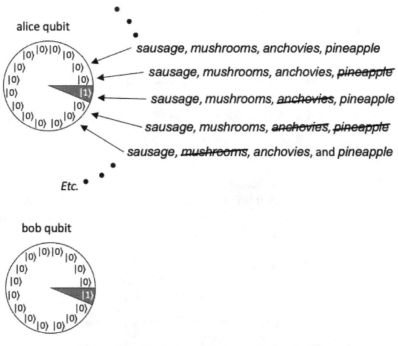

Figure 8.31 – The alice and bob qubits visualized

A good scratch variable always starts off being zero – not some superposition of $|1\rangle$ and $|0\rangle$. So, before moving on to the next step in Grover's algorithm, we must return the **alice** and **bob** qubits to their original pristine zero states. This means changing the parts of those qubits' superposition back to $|0\rangle$. We must undo what was done by the two tall control gates in *Figure 8.30*. Since control gates of this kind are their own inverses, we can undo these gates by applying them a second time. This is called **uncomputing**, and it's illustrated in *Figure 8.32*:

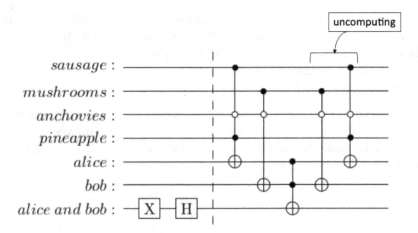

Figure 8.32 – An oracle for a delicious pizza

That's how we create an oracle for a SAT problem.

Gates for the diffuser

Compared with the complexities of designing a circuit for an oracle, designing a diffuser circuit is easy. The diffuser doesn't change much from one search problem to another. To create a diffuser for a particular problem, all you need to know is the number of qubits involved. For the pizza problem in the previous section, the diffuser looks like the one in *Figure 8.33*:

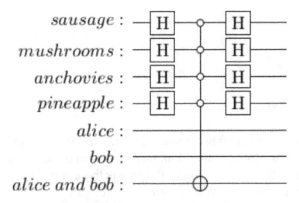

Figure 8.33 – The diffuser for a four-qubit Grover search

In this circuit, the multi-qubit control gate flips the $|0000\rangle$ amplitude, which is exactly what we want. To see why, revisit the *A matrix for the diffuser* section.

Coding Grover's algorithm with quantum gates

Let's modify the code from the *Coding Grover's algorithm with matrices* section and use quantum gates instead.

In your Qiskit code, you frequently call `circ.x(0)`. Behind the scenes, the x method's code imports a class named XGate from its `standard_gates` package. To create an oracle's big controlled NOT gates, we need to explicitly import that XGate class:

```
from qiskit.circuit.library.standard_gates import XGate
```

Here's the code to create an oracle to find the $|100\rangle$ amplitude:

```
oracle = QuantumCircuit(4)
oracle.x(3)
oracle.h(3)

ctrl = XGate().control(3, ctrl_state='100')
oracle.append(ctrl, qargs=[0, 1, 2, 3])

oracle.barrier()
display(oracle.draw('latex'))
```

In the code for the oracle, `XGate().control(3, ctrl_state='100')` tells Qiskit to control an *X* gate using 3 control qubits. The `'100'` string describes the sequence of control values required to activate the *X* gate. A subsequent call to append pastes this new controlled *X* structure onto the existing `oracle` circuit . (See *Figure 8.34*.)

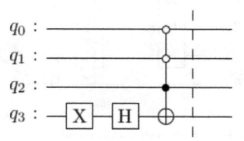

Figure 8.34 – A three-qubit oracle

In the call to append, the `qargs` parameter is similar to the `qubits` parameter in an earlier example's `unitary` method. The `qargs` list defines which of the `ctrl` gate's qubits match with which of the `oracle` circuit's qubits. A different list, `qargs=[1, 2, 3, 0]`, would create the circuit shown in *Figure 8.35*:

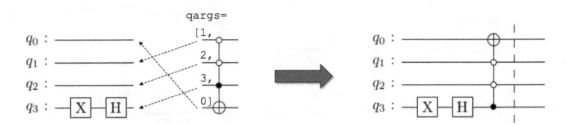

Figure 8.35 – Matching a control gate's qubits with an existing circuit's qubits

In the code for the diffuser, the control qubits flip the circuit's $|000\rangle$ amplitude:

```
diffuser = QuantumCircuit(4)
diffuser.h([0, 1, 2])

ctrl = XGate().control(3, ctrl_state='000')
diffuser.append(ctrl, qargs=[0, 1, 2, 3])

diffuser.h([0, 1, 2])

diffuser.barrier()
display(diffuser.draw('latex'))
```

This code creates a circuit, as shown in *Figure 8.36*:

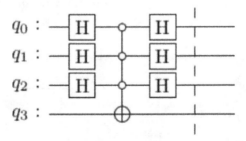

Figure 8.36 – A diffuser made from a multi-qubit control gate

You can substitute this section's `oracle` and `diffuser` instructions for the corresponding code in the *Coding Grover's algorithm with matrices* section. When you do, you get yet another implementation of Grover's algorithm in Qiskit.

Epilogue – what does π have to do with Grover's algorithm?

When you run Grover's algorithm, the optimal number of iterations is $\frac{\pi}{4}\sqrt{N}$, where N is the number of things you're searching through. In this formula, you have n qubits and $N = 2^n$.

Usually, the formula requires a few tweaks. For example, you can't use the formula to search through exactly 1,000 items, because 1,000 isn't a power of 2. Instead, you add 24 fake items to your list. Then, you use 10 qubits to search through $2^{10} = 1024$ items. And what if $\frac{\pi}{4}\sqrt{N}$ isn't an integer? The value of $\frac{\pi}{4}\sqrt{1024}$ is approximately 25.1327, and you can't apply the Grover iterate 25.1327 times. In this case, there are ways to decide whether Grover's sweet spot is 25 or 26, but we won't get into that here.

In this section, our goal is to convey some idea of the role $\frac{\pi}{4}$ plays in the number of Grover iterate applications. Our explanation has many gaps and involves several approximations. We won't come to any undeniable conclusions. But, by reading this, you'll add a piece of π to the coffee you had when you searched through boxes in previous sections.

Figure 8.37 pictures our qubits' amplitudes as axes on a plane:

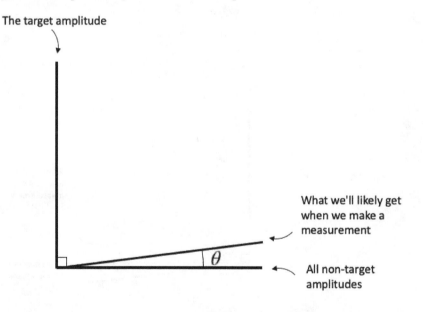

Figure 8.37 – A different way to think about Grover's algorithm

The little square in the diagram's lower left corner reminds us that the "target" and "non-target" lines are perpendicular to one another. The angle between them is $\frac{\pi}{2}$. As in the previous sections, we assume that the non-target amplitudes dominate. In *Figure 8.37*, I made the θ angle look like it's about 7 degrees, but that's only because I can't squeeze θ between two lines that are only 0.000000000001 degrees apart.

When we apply the Grover iterate, we add 2θ to this preceding figure . (See *Figure 8.38*.)

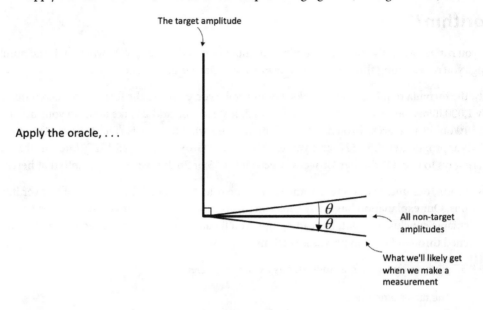

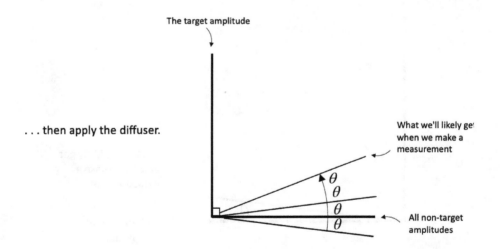

Figure 8.38 – One application of the Grover iterate

The amplitude changes in *Figure 8.38* correspond exactly to the two-distance addition between *Figures 8.13* and *8.14*. The difference between the non-target amplitudes and the likely outcome of a measurement is now 3θ away from the non-target amplitudes.

Let's apply the Grover iterate again. The result is shown in *Figure 8.39*:

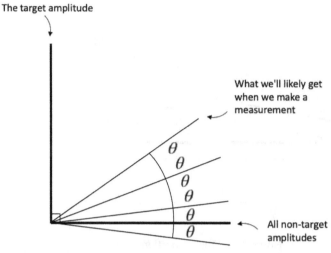

Figure 8.39 – After two applications of the Grover iterate

Just as it was in *Figure 8.18*, we now have a multiple of 5. The angle between the non-target amplitudes and the likely measurement outcome is 5θ away from the non-target amplitudes.

So, when does this process reach its optimal level? After k applications of the Grover iterate, the angle between the "likely measurement" and "non-target" lines is $(2k + 1)\theta$. We want this angle to be $\frac{\pi}{2}$ because that makes the "likely measurement" line touch the "target" line:

$$(2k + 1)\theta = \frac{\pi}{2}$$

$$2k\theta \approx \frac{\pi}{2}$$

$$k \approx \frac{\pi}{4} \cdot \frac{1}{\theta}$$

If I tell you why $\frac{1}{\theta} \approx \sqrt{N}$, we're done! Look at *Figure 8.40*:

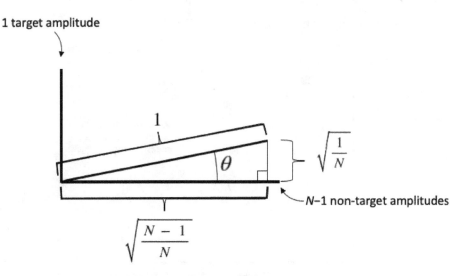

Figure 8.40 – The proportions of a triangle

The N and N – 1 values play a significant role in the run of Grover's algorithm. These values are naturally part of the triangle formed by the θ angle. To check that we're on the right track in *Figure 8.40*, notice that the three line lengths obey the Pythagorean theorem:

$$\left(\sqrt{\frac{N-1}{N}}\right)^2 + \left(\sqrt{\frac{1}{N}}\right)^2 = 1^2$$

Dividing the triangle's height by its hypotenuse, we get $\sin\theta = \sqrt{\frac{1}{N}}$. But remember that N is normally very large, so the θ angle is very small. And, for very small values of θ, we know that $\sin\theta = \theta$. So, here's what we have:

$$\sin\theta = \sqrt{\frac{1}{N}}$$

$$\theta \approx \sqrt{\frac{1}{N}}$$

$$\frac{1}{\theta} \approx \sqrt{N}$$

That's how the optimal number of iterations comes to be $\frac{\pi}{4}\sqrt{N}$.

Summary

Grover's algorithm speeds up the search of an unordered list. We represent a list of size N with n qubits, where $N = 2^n$. Eventually, when we measure the qubits, we see a combination of n bits. Each possible combination stands for an element in the list. Each step of Grover's algorithm increases the probability that the measurement outcome represents the target of our search.

The optimal number of steps depends on the number of elements in our unordered list. Each step of Grover's algorithm makes approximately the same number increase in the target combination's amplitude. Since an amplitude is the square root of a probability, the optimal number of steps grows with $\sqrt{N}$. That's better than a classical search, where the optimal number of steps grows with N.

Grover's search can be useful when it's easy to verify that a particular element is the search target but difficult to find the search target among all the choices. Problems that have this characteristic include password encryption, optimization, and Boolean satisfiability.

In the next chapter, we will push the practicality boundary even further with Shor's algorithm. The algorithm applies to a critical problem – RSA encryption. It could reduce the time to crack RSA from trillions of years to a few hours or a few seconds.

Questions

1. How many qubits do we need to search for something in a list containing 1,000,000 entries? How many steps will the search take?

2. Write the matrix of the diffuser to search through an eight-value list.

3. Modify the code in this chapter's *Coding Grover's algorithm with matrices* section so that the target is a randomly chosen state between $|000\rangle$ and $|111\rangle$.

4. For what values of x, y, and z is the following expression true?

    ```
    (x | y | ~z) & (x | ~y | z) & (x | ~y | ~z) &
    (~x | y | z) & (~x | y | ~z) & (~x | ~y | z) &
    (~x | ~y | ~z)
    ```

    Legend: | means or, ~ means not, and & means and.

5. Run the following code:

    ```
    from qiskit import QuantumCircuit, execute
    from qiskit import Aer
    from qiskit.visualization import array_to_latex

    circ = QuantumCircuit(3)
    circ.x(2)
    circ.h(2)
    circ.barrier()
    ```

```
circ.toffoli(0, 1, 2)
display(circ.draw('latex'))
device = Aer.get_backend('unitary_simulator')
job = execute(circ,backend = device,shots = 1)
matrix = job.result().get_unitary()
array_to_latex(matrix)
```

What part of the output's run verifies that the circuit is an oracle for the $|11\rangle$ state?

6. Consider the following scenario:

    Alice wants to see *The Identity Matrix* or *The Sixth Tensor*. Bob wants to see *The Sixth Tensor* or *The Jupyter Notebook*. (Alice doesn't want to see *The Jupyter Notebook*, and Bob doesn't want to see *The Identity Matrix*.)

    Alice and Bob won't know which movie to see until they hear from a run of Grover's algorithm. Write code for Alice and Bob using Qiskit's `PhaseOracle` and `Grover` classes.

7. According to De Morgan's laws, we have the following:

    ```
    not (p or q)  = (not p) and (not q)
    not (p and q) = (not p) or  (not q)
    ```

    Use De Morgan's laws to eliminate any use of the `or` operator from the Boolean expression in *Question 6*.

    Draw an oracle for your Boolean expression using the techniques in the *Gates and circuits for Grover's algorithm* section.

9
Shor's Algorithm

Much of the buzz around quantum computing comes from one simple fact: a quantum computer with several thousand qubits can solve a certain strategic problem that classical computers have no hope of solving.

In 1994 [1], Peter Shor unveiled an algorithm to crack today's widely used encryption schemes. Decrypting a message might take trillions of years on a classical computer. But Shor's algorithm, running on a sufficiently large quantum computer, can decrypt a message in less than a minute. It would be nice if most people welcomed a solution to this decryption problem. But, unfortunately, most people who want to break encryption schemes are malicious hackers.

Businesses and governments are taking this problem seriously. At this very moment (no matter when you're reading this), people around the world are developing **post-quantum cryptography** schemes. These newly formulated schemes must be resistant to vulnerabilities arising from quantum computing.

Shor's algorithm involves lots of math, so this chapter's focus differs from that of previous chapters. In this chapter, we will present examples involving ridiculously small numbers to show you the highlights of Shor's algorithm. We won't cover all the nooks and crannies within Shor's algorithm, and, in this regard, we're not alone. Many technical books omit details of Shor's algorithm.

We'll cover the following topics in this chapter:

- A popular encryption scheme
- How Shor's algorithm works
- Complex numbers
- Finding a sequence's period
- Shoring up your knowledge
- Illustrating Shor's algorithm with Qiskit code

Technical requirements

You can find the code used in this chapter on GitHub:

`https://github.com/PacktPublishing/Quantum-Computing-Algorithms`

A popular encryption scheme

In August 1977, three researchers wrote an article for a mathematical curiosities column in the *Scientific American* magazine [2]. The article described what has come to be known as **RSA encryption**, so named after its originators—Rivest, Shamir, and Adelman. The idea behind RSA is that multiplying numbers is easy, but factoring numbers is difficult. In *Figure 9.1*, we get a 100-digit number by multiplying two 50-digit numbers:

```
                37975227936943673922808872755445627854565536638199
              × 40094690950920881030683735292761468389214899724061
                ──────────────────────────────────────────────────
                37975227936943673922808872755445627854565536638199
              227851367621662043536853236532673767127393219829194
            1519009117477746956912354910217825114182621465527960
                                    •
                                  •
                                •
      1519009117477746956912354910217825114182621465527960 0
      ──────────────────────────────────────────────────────────────
      152260502792253336053561837813263742971806811496138068865790849458012296325895289765 4000350692006139
```

Figure 9.1 – The RSA-100 number

When I presented this multiplication problem to my laptop computer, I got the answer almost instantly. Multiplying two numbers, however large they may be, isn't challenging for today's hardware.

But what if we try to solve the problem in reverse? What if we start with the 100-digit number at the bottom of *Figure 9.1*, and ask a computer to find the two 50-digit numbers at the top of the figure? When I handed this task to a respected mathematics website, the site replied with a "*time exceeded*" message. I didn't run the problem on my own laptop for fear of starting a house fire.

The general flow of RSA encryption is shown in *Figure 9.2*:

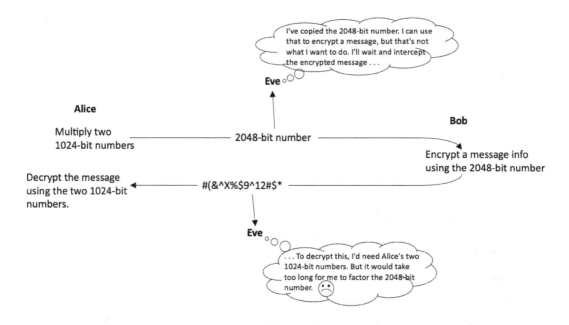

Figure 9.2 – RSA encryption overview

In *Figure 9.2*, Alice randomly generates two 1024-bit numbers. These two numbers form Alice's **private key**. Alice multiplies the two 1024-bit numbers together (which is easy to do), and sends the 2048-bit product to Bob. This 2048-bit product is the scheme's **public key**. Anyone can have access to the public key—even the malicious agent Eve.

Using the public key, anyone can encrypt a message. But once Bob uses the public key to encrypt the message, no one can decrypt Bob's mumbo-jumbo without having the private key. The math behind RSA encryption almost guarantees this.

Getting the private key from the public key would require factoring the public key. And factoring the public key would require many millennia on a classical computer.

With RSA encryption, the public key doesn't have to be kept secret. Anyone can have the public key. Because of this, RSA is an example of **public key encryption**, also known as **asymmetric encryption**.

You may be wondering what Alice and Bob do in order to encrypt and decrypt a message. The math involved in RSA encryption is the topic of the next section.

An example of RSA encryption

In this section, we will use RSA to encrypt text and then decrypt the encrypted text. We will show you *how* RSA works, but not *why* RSA works. If you're not curious about the nuts and bolts of RSA, you can skip this section.

As a first step in RSA encryption, Alice randomly generates two very large prime numbers. To make this section readable, we'll use two small numbers: 5 and 11. Both numbers are prime. (That's important.) The product of these two numbers is 55. So, in this tiny example, the private key contains the numbers 5 and 11, and the public key contains 55. Of course, a classical computer can factor 55 in nanoseconds. So, in practice, Alice would never use such small numbers to implement RSA encryption.

As a next step, Alice computes the **totient** of 5 and 11. Here's how she does it:

$$\text{totient} = (5 - 1) (11 - 1) = 40$$

As if Alice doesn't have enough numbers to remember, Alice then selects two additional numbers. One of them is her coprime number. To be called a **coprime**, a number must be less than the totient and have no divisors (other than 1) in common with the totient. For example, with a totient of 40, the coprime can't be 5, because 5 divides evenly into 40. Nor can the coprime be 6 because 2 divides evenly into both 40 and 6.

In this example, let's assume that Alice selects the coprime 7. The numbers 40 and 7 have no common divisors other than 1.

Alice must also find an **inverse coprime**—a number that satisfies the equation—as follows:

$$(\text{coprime} \cdot \text{inverse\_coprime}) \ \% \ \text{totient} = 1$$

In this equation, % stands for Python's modulo operator. For an inverse coprime, Alice finds the number 23:

$$7 \cdot 23 \ \% \ 40 = 161 \ \% \ 40 = 1$$

Table 9.1 shows what we have so far:

Private key	5 and 11
Public key	55
Totient	40
Coprime	7
Inverse coprime	23

Table 9.1 – Alice's RSA values

At last, Alice can send a public key to Bob! In fact, Alice sends both the public key, 55, and the coprime, 7.

Bob wants to encrypt a very short message. His message is the number 9. Alice might understand this to mean "The ASCII code for the horizontal tab" or "Deploy the SWAT team at 0900." What the number 9 means to Alice and Bob is none of our concern.

Anyway, with 9 as the plaintext, Bob encrypts this value using the following formula:

$$\text{plaintext}^{\text{coprime}} \ \% \ \text{public\_key} = \text{ciphertext}$$
$$9^7 \quad \% \quad 55 \quad = 4$$

Bob sends the number 4 back to Alice. When Alice receives the number 4, she decrypts that value using the following formula:

$$\text{ciphertext}^{\text{inverse\_coprime}} \ \% \ \text{public\_key} = \text{plaintext}$$
$$4^{23} \quad \% \quad 55 \quad = 9$$

You may notice that Alice doesn't actually use the private key numbers, 5 and 11, to decrypt the message. That doesn't matter. Alice used those private numbers to find the inverse coprime, and that inverse coprime must be kept from Eve's view.

> **Important note**
>
> We'll use the term *coprime* in the next section to describe a sequence of numbers that forms the basis for Shor's algorithm. When you read that section, remember not to confuse Shor's coprime with this section's RSA scheme's coprime. The two coprimes are related to one another because they both have the same mathematical underpinnings, but these two numbers play different roles in RSA encryption and Shor's algorithm.

How Shor's algorithm works

If you remember only one thing about the strategy behind Shor's algorithm, it should be this: the algorithm examines the repetition within a particular sequence of numbers and uses that repetition to factor the public key.

Let's take a look at some sequences of numbers. (Once again, the numbers in our examples are laughingly small to ensure that the arithmetic is manageable.) On a rainy day in August, Alice initiates a sensitive conversation with Bob. While Eve does her snooping, she notices Alice sending the public key 15 to Bob. Eve's goal is to factor 15 into its component parts.

With the help of her computer, Eve chooses a number that's smaller than 15. As we did in the previous section, we'll call this smaller number a **coprime**. As with the previous section's coprime, this smaller number must have no divisors (other than 1) in common with the public key 15.

In this example, let's have Eve select the coprime number 2.

Eve methodically computes the powers of the coprime, divides those powers by 15, and makes note of the remainder:

$$2^0 \ \% \ 15 \ = \ 1 \ \% \ 15 \ = \ 1$$
$$2^1 \ \% \ 15 \ = \ 2 \ \% \ 15 \ = \ 2$$
$$2^2 \ \% \ 15 \ = \ 4 \ \% \ 15 \ = \ 4$$
$$2^3 \ \% \ 15 \ = \ 8 \ \% \ 15 \ = \ 8$$
$$2^4 \ \% \ 15 \ = \ 16 \ \% \ 15 \ = \ 1$$
$$2^5 \ \% \ 15 \ = \ 32 \ \% \ 15 \ = \ 2$$
$$2^6 \ \% \ 15 \ = \ 64 \ \% \ 15 \ = \ 4$$
$$2^7 \ \% \ 15 \ = \ 128 \ \% \ 15 \ = \ 8$$
$$2^8 \ \% \ 15 \ = \ 256 \ \% \ 15 \ = \ 1$$
$$2^9 \ \% \ 15 \ = \ 512 \ \% \ 15 \ = \ 2$$
$$2^{10} \ \% \ 15 \ = \ 1024 \ \% \ 15 \ = \ 4$$

Etc.

Let's call 1, 2, 4, 8, 1, 2, 4, 8, 1, 2, 4, ... a **coprime powers sequence**.

Eve observes that the coprime powers sequence contains a repeating pattern consisting of four numbers: 1, 2, 4, and 8. The fact that it's exactly four numbers is important, so Eve records the number 4 as the pattern's **period**. Eve does some arithmetic and finds the following numbers:

coprime$^{period/2}$ – 1	coprime$^{period/2}$ + 1
$2^{4/2}$ – 1	$2^{4/2}$ + 1
3	5

And there they are! Eve has the numbers 3 and 5. She's factored the public key and discovered Alice's private key. Eve can use these numbers to decrypt Bob's message.

What happens when Eve selects 7 (instead of 2) as her coprime? Here's the coprime powers sequence that she gets:

$$7^0 \% 15 = 1 \% 15 = 1$$
$$7^1 \% 15 = 7 \% 15 = 7$$
$$7^2 \% 15 = 49 \% 15 = 4$$
$$7^3 \% 15 = 343 \% 15 = 13$$
$$7^4 \% 15 = 2401 \% 15 = 1$$
$$7^5 \% 15 = 16807 \% 15 = 7$$
$$7^6 \% 15 = 117649 \% 15 = 4$$
$$7^7 \% 15 = 823543 \% 15 = 13$$
$$7^8 \% 15 = 5764801\% 15 = 1$$

Etc.

Once again, Eve finds a period of 4. (That's a coincidence. The period isn't always 4.) But this time, Eve's arithmetic yields some different numbers:

coprime$^{period/2}$ – 1	coprime$^{period/2}$ + 1
$7^{4/2}$ – 1	$7^{4/2}$ + 1
48	50
2·2·2·2·3	2·5·5

Factors 3 and 5 are embedded into the results that Eve obtains. Now, Eve has three numbers (48, 50, and 15) containing the public key's factors. Eve can use some other mathematics tricks to compare these three numbers and discover that 3·5 = 15. She can use this information to decrypt Bob's message.

Why does the period of a sequence tell Eve so much about factoring the number 15? To find out, keep reading.

The role of a period in factoring a number

In the previous section, we attacked the public key 15 with a coprime of 7. In the end, we came up with the numbers $7^{4/2}$ – 1 and $7^{4/2}$ + 1. If we multiply these together, we get the following:

$$(7^{4/2} - 1)(7^{4/2} + 1) = 7^{4/2}7^{4/2} - 7^{4/2} + 7^{4/2} - 1 = 7^4 - 1$$

But when we created our coprime powers sequence, we found that $7^4 \% 15 = 1$. So, here's our line of reasoning:

- When we divide 7^4 by 15, we get a remainder of 1

- Therefore, when we divide 7^4 – 1 by 15, we get a remainder of 0

- That is, 15 divides evenly into $7^4 - 1$
- Therefore, 15 divides evenly into $(7^{4/2} - 1)(7^{4/2} + 1)$

In fact, when we divide 15 into $(7^{4/2} - 1)(7^{4/2} + 1)$, we get the following:

$$\frac{(2 \cdot 2 \cdot 2 \cdot 2 \cdot 3)(2 \cdot 5 \cdot 5)}{3 \cdot 5}$$

In this expression, notice how the numbers 3 and 5 occur in both the numerator and the denominator. That's not an accident. Since 3 and 5 occur in the denominator, they must occur somewhere in the numerator. After all, 3 and 5 are prime numbers. You can't factor 3 into two smaller integers. The same is true of the number 5. In a sense, each prime number is an atom.

Since 3 divides evenly into $(7^{4/2} - 1)(7^{4/2} + 1)$, we have to be able to cancel 3 with $(7^{4/2} - 1)(7^{4/2} + 1)$. But, with 3 being atomic, we can't find part of 3 in $(7^{4/2} - 1)$ and another part of 3 in $(7^{4/2} + 1)$. We must find all of 3 in either $(7^{4/2} - 1)$ or $(7^{4/2} + 1)$. See my fanciful illustration in *Figure 9.3*:

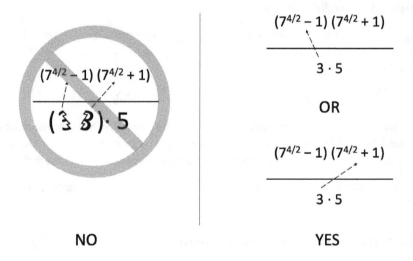

Figure 9.3 – You can't split the atom

In the same way, we must find all of 5 in either $(7^{4/2} - 1)$ or $(7^{4/2} + 1)$. So, the factors of 15 are embedded into the numbers that Eve finds when she discovers period 4 in the coprime powers sequence. By running the well-known **Euclidean algorithm** on a classical computer, Eve can find the greatest common divisors $\gcd(15, 7^{4/2} - 1)$ and $\gcd(15, 7^{4/2} + 1)$.

Suppose that Eve starts by finding $\gcd(15, 7^{4/2} - 1) = \gcd(15, 48) = 3$. Ordinary division is easy, even when very large numbers are involved. With this tiny toy example, Eve divides 3 into 15 and gets 5. *Et voilà!* Eve has the two factors of Alice's key. Eve can steal Bob's information!

For Eve, the only challenging part is finding the period of the coprime powers sequence. With the two examples in the *How Shor's algorithm works* section, that period was 4. But in general, the period isn't 4. With a very large public key, Eve's classical computer would take millennia trying to find the period. What Eve needs is a way for a quantum computer to find the period quickly and efficiently. Fortunately for Eve (but unfortunately for Alice and Bob), such a way exists. It's called the **Quantum Fourier Transform (QFT)**, and it's the subject of several sections in this chapter.

Repeated squaring

In the previous sections, we find $7^n \% 15$ up to the point where $n = 8$. The numbers that are involved become fairly hefty. When n is 8, we're asked to do arithmetic on a seven-digit number. In this example, the so-called public key is 15. Imagine what happens in a real-life example when the public key is a 2048-bit number!

To remedy this situation, we rely on **repeated squaring**. To illustrate this technique, let's divide powers of 13 by 22. First, we'll do it using the brute-force approach:

$$13^0 \% 22 = 1$$

$$13^1 \% 22 = 13$$

$$13^2 \% 22 = 169 \% 22 = 15$$

$$13^3 \% 22 = 2197 \% 22 = 19$$

$$13^4 \% 22 = 28561 \% 22 = 5$$

$$13^5 \% 22 = 371293 \% 22 = 21$$

$$13^6 \% 22 = 4826809 \% 22 = 9$$

$$13^7 \% 22 = 62748517 \% 22 = 7$$

$$13^8 \% 22 = 815730721 \% 22 = 3$$

By the time we calculate 13^8, we haven't started repeating values, and we're already dealing with a nine-digit number! There has to be a better way.

The idea behind repeated squaring is to chop each number into parts and apply the % operator to each part. For example, $13^8 = 13^4 13^4$. Since $13^4 \% 22 = 5$, we can find $13^8 \% 22$ by taking $5 \cdot 5 \% 22$. That's $25 \% 22$, which is 3. With a little cleverness, we've gone from having to find $815730721 \% 22$ to simply finding $25 \% 22$. That's a big improvement!

We can use repeated squaring throughout the calculation of a coprime powers sequence. Here's how it works for the powers 13^0 to 13^8:

$$13^0 \ \% \ 22 = 1$$

$$13^1 \ \% \ 22 = 13$$

$$13^2 \ \% \ 22 = 169 \ \% \ 22 = 15$$

$$13^3 \ \% \ 22 = \left(13^2 \ \% \ 22\right)\left(13^1 \ \% \ 22\right) \ \% \ 22 = 15 \cdot 13 \ \% \ 22 = 195 \ \% \ 22 = 19$$

$$13^4 \ \% \ 22 = \left(13^2 \ \% \ 22\right)\left(13^2 \ \% \ 22\right) \ \% \ 22 = 15 \cdot 15 \ \% \ 22 = 225 \ \% \ 22 = 5$$

$$13^5 \ \% \ 22 = \left(13^4 \ \% \ 22\right)\left(13^1 \ \% \ 22\right) \ \% \ 22 = 5 \cdot 13 \ \% \ 22 = 65 \ \% \ 22 = 21$$

$$13^6 \ \% \ 22 = \left(13^4 \ \% \ 22\right)\left(13^2 \ \% \ 22\right) \ \% \ 22 = 5 \cdot 15 \ \% \ 22 = 75 \ \% \ 22 = 9$$

$$13^7 \ \% \ 22 = \left(13^4 \ \% \ 22\right)\left(13^2 \ \% \ 22\right)\left(13^1 \ \% \ 22\right) \ \% \ 22 = 5 \cdot 15 \cdot 13 \ \% \ 22 = 975 \ \% \ 22 = 7$$

$$13^8 \ \% \ 22 = \left(13^4 \ \% \ 22\right)\left(13^4 \ \% \ 22\right) \ \% \ 22 = 5 \cdot 5 \ \% \ 22 = 25 \ \% \ 22 = 3$$

At no point in this new process did we deal with a number larger than three decimal digits. Nice work!

Every number in a coprime powers sequence is a real number with no imaginary part. But to find the period of a coprime powers sequence, Shor's algorithm makes ample use of complex numbers. So, in the next section, we will explore some properties of the complex number system.

Complex numbers

Much of the work in Shor's algorithm depends on a matrix known as the QFT. Using the QFT, Eve can find the period of a sequence and use that period to decrypt Bob's message. The QFT uses complex numbers to perform its magic, so this section covers some facts about complex numbers.

Complex number basics

Imaginary numbers were first described by Girolamo Cardano in 1545. The best-known imaginary number is $\sqrt{-1}$, also known as i, and sometimes in Python code, j. The worst part of imaginary numbers is their name. If we called them "super numbers" instead of "imaginary numbers," people wouldn't be so suspicious of them. It's true that imaginary numbers don't arise naturally in day-to-day situations. So, for most people, imaginary numbers don't exist. But scientists rely on imaginary numbers all the time. Imaginary numbers are very useful.

A **complex number** is a number with two parts. One of those parts isn't imaginary, and the other part is. Take, for example, the number $3 + 2i$. It has the **real part** 3, and the **imaginary part** $2i$. You can describe this number visually on the graph in *Figure 9.4*:

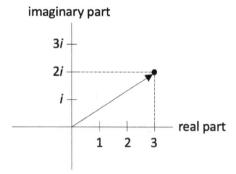

Figure 9.4 – The number 3 + 2*i*

To add complex numbers, you add the real part to the real part and add the imaginary part to the imaginary part:

$$\begin{array}{r} 3 + \ 2i \\ 7 + \ 9i \\ \hline + \ 10 + 11i \end{array}$$

Subtraction works the same way. To find the product of two complex numbers, you cross-multiply, like so:

$$(3 + 2i)(7 + 9i)$$

$$= 3 \cdot 7 + 2i \cdot 7 + 3 \cdot 9i + 2i \cdot 9i$$

$$= 21 + 14i + 27i + 18\sqrt{-1}^{2}$$

$$= 21 + 41i - 18$$

$$= 3 + 41i$$

Chapter 3 makes the following claim: When you measure a qubit $\begin{pmatrix} x \\ y \end{pmatrix}$, the probability of getting 0 is $|x|^2$, and the probability of getting 1 is $|y|^2$. Until this point in the book, we've ignored the absolute value signs because, for a real number, r, there's no difference between r^2 and $|r|^2$. But complex numbers beat to a different drum. For a complex number, the absolute value is the distance between that number and $0 + 0i$. (See *Figure 9.5*.)

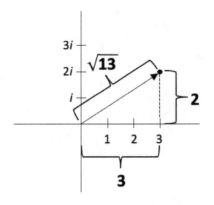

$$|3 + 2i| = \sqrt{3^2 + 2^2} = \sqrt{13}$$

Figure 9.5 – The absolute value of a complex number

The vector $\begin{pmatrix} \frac{1}{2} - \frac{i}{2\sqrt{3}} \\ \frac{1}{\sqrt{2}} + \frac{i}{\sqrt{6}} \end{pmatrix}$ represents a qubit because, when you add up the appropriate numbers, you get 1.

Here's the math:

$$\left| \frac{1}{2} - \frac{i}{2\sqrt{3}} \right|^2 + \left| \frac{1}{\sqrt{2}} + \frac{i}{\sqrt{6}} \right|^2$$

$$= \left(\sqrt{\left(\frac{1}{2}\right)^2 + \left(-\frac{1}{2\sqrt{3}}\right)^2} \right)^2 + \left(\sqrt{\left(\frac{1}{\sqrt{2}}\right)^2 + \left(\frac{1}{\sqrt{6}}\right)^2} \right)^2$$

$$= \left(\frac{1}{2}\right)^2 + \left(-\frac{1}{2\sqrt{3}}\right)^2 + \left(\frac{1}{\sqrt{2}}\right)^2 + \left(\frac{1}{\sqrt{6}}\right)^2$$

$$= \frac{1}{4} + \frac{1}{12} + \frac{1}{2} + \frac{1}{6} = 1$$

Let's look at unitary matrices next.

Unitary matrices

We introduced unitary matrices in *Chapter 3*. A matrix that contains *only real numbers* is called a **unitary matrix** as long as the inverse of the matrix is the matrix's transpose. Unitary matrices are important because those are the matrices that can be implemented in a quantum circuit.

With the introduction of complex numbers, we must refine our definition of a unitary matrix. We start by defining a complex conjugate. (See *Figure 9.6*.)

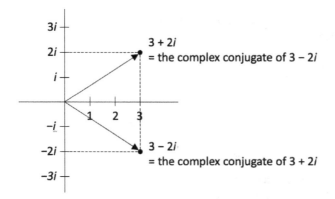

Figure 9.6 – Complex conjugate

In general, the complex conjugate of $a + bi$ is $a - bi$. We use an asterisk to denote a complex conjugate. So, for example, $(3 + 2i)^*$ stands for the number $3 - 2i$.

Why do we bother to define complex conjugates? Consider the following matrix:

$$S = \begin{pmatrix} 1 & 0 \\ 0 & i \end{pmatrix}$$

If we take the transpose of S, we don't get the inverse of S:

$$S^T S = \begin{pmatrix} 1 & 0 \\ 0 & i \end{pmatrix}\begin{pmatrix} 1 & 0 \\ 0 & i \end{pmatrix} = \begin{pmatrix} 1 & 0 \\ 0 & i^2 \end{pmatrix} = \begin{pmatrix} 1 & 0 \\ 0 & -1 \end{pmatrix} \neq I$$

In the matrix S, the value i isn't a real number. So, to get the inverse S^{-1}, we find the transpose of the matrix and then take each entry's complex conjugate. This gives us what's called the **adjoint** of the matrix. The symbol for the adjoint is a superscript dagger ($\dagger$):

$$S^{-1} = S^\dagger = \begin{pmatrix} 1^* & 0^* \\ 0^* & i^* \end{pmatrix} = \begin{pmatrix} 1 & 0 \\ 0 & -i \end{pmatrix}$$

$$S^\dagger S = \begin{pmatrix} 1 & 0 \\ 0 & -i \end{pmatrix}\begin{pmatrix} 1 & 0 \\ 0 & i \end{pmatrix} = \begin{pmatrix} 1 & 0 \\ 0 & -i^2 \end{pmatrix} = \begin{pmatrix} 1 & 0 \\ 0 & 1 \end{pmatrix} = I$$

In the preceding equations, notice that $1^* = 1$ and $0^* = 0$. Taking the complex conjugate of a real number has no effect. That's why we could ignore complex conjugates before this point in the book. Until this chapter, the entries in our matrices were all real numbers.

For any matrix containing complex numbers, the matrix is **unitary** as long as its adjoint is the matrix's inverse. Therefore, the S matrix is unitary. Unitary matrices unitary matrix are exactly the ones that can be implemented as quantum gates. So quantum computers have S gates. Qiskit's s function adds an S gate to a circuit. Here's some code to prove it:

```
from qiskit import QuantumCircuit
from qiskit.visualization import visualize_transition

circ = QuantumCircuit(1)
circ.h(0)
circ.s(0)

display(circ.draw('latex', scale=2.5))
visualize_transition(circ, trace=True)
```

In *Figure 9.7*, I've modified the code's output to highlight the effects of the H and S gates:

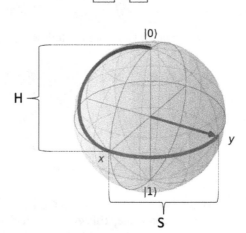

Figure 9.7 – Applying H and then S

An **S gate** rotates a qubit one-quarter of the way around the Bloch sphere's *z* axis. So, in *Figure 9.7*, the S gate takes the qubit in an arc one-quarter of the way around the equator. The S gate's transition goes from $\frac{1}{\sqrt{2}}\vee 0\rangle + \frac{1}{\sqrt{2}}\vee 1\rangle$ to $\frac{1}{\sqrt{2}}\vee 0\rangle + i\frac{1}{\sqrt{2}}\vee 1\rangle$. That makes sense because the following applies:

$$S\begin{pmatrix}\dfrac{1}{\sqrt{2}}\\[2mm]\dfrac{1}{\sqrt{2}}\end{pmatrix} = \begin{pmatrix}1 & 0\\0 & i\end{pmatrix}\begin{pmatrix}\dfrac{1}{\sqrt{2}}\\[2mm]\dfrac{1}{\sqrt{2}}\end{pmatrix} = \begin{pmatrix}\dfrac{1}{\sqrt{2}}\\[2mm]i\dfrac{1}{\sqrt{2}}\end{pmatrix}$$

What happens in *Figure 9.7* when you apply an S gate three more times? You continue your way around the Bloch sphere's equator and end up back at $\frac{1}{\sqrt{2}}$ v 0⟩ + $\frac{1}{\sqrt{2}}$ v 1⟩. In the next several sections, we will explore this phenomenon and its consequences.

The connection between complex numbers and circles

Going around in a circle means repeating things over and over again, which is what we do when we build a coprime powers sequence. So, the circular nature of certain complex numbers plays an important role in Shor's algorithm. In *Figure 9.8*, we graph four complex numbers:

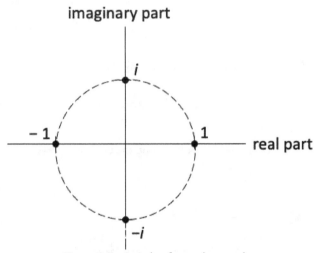

Figure 9.8 – A circle of complex numbers

If it helps, think of the circle in *Figure 9.8* as the equator of the Bloch sphere, with 1 coming at you out of the page and −1 at the back of the sphere trying to hide from you.

The four numbers 1, i, −1, and −i are known as the **fourth roots of unity** because they're the four solutions to the equation $x^4 = 1$. Of course, there's nothing special about the number 4. In *Figure 9.9*, we graph the **eighth roots of unity**:

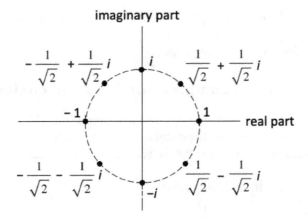

Figure 9.9 – The eighth roots of unity

If the numbers in *Figure 9.9* look burdensome, don't despair. There are many ways to label the roots of unity. We list a few of them here:

- With the fourth roots of unity, the numbers around the circle are powers of *i*. (See *Figure 9.10*.)

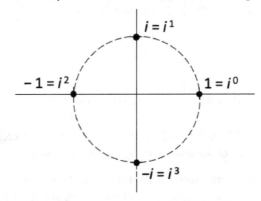

Figure 9.10 – Powers of $\sqrt{-1}$

The same kind of thing works with any collection of roots of unity. For example, if you define ω to be $\frac{1}{\sqrt{2}} + \frac{1}{\sqrt{2}}i$, then you get the eighth roots of unity shown in *Figure 9.11*:

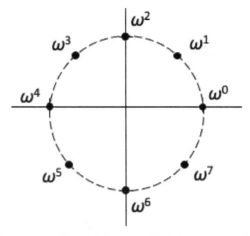

Figure 9.11 – The eighth roots of unity as powers of ω

Figures 9.9 and *9.11* contain the same eight numbers. The only difference is the way those numbers are represented in the two figures. In fact, it's a common practice to represent the roots of unity as powers of ω no matter how many roots you have. We'll use this notation throughout the chapter.

- I sometimes find it useful to represent the roots of unity with little circles. *Figure 9.12* shows you what I mean:

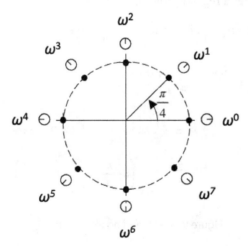

Figure 9.12 – Circle notation for the eighth roots of unity

With this unconventional notation, it's easy to see how the roots of unity relate to one another. For example, the fact that ⊘ and ⊘ point in opposite directions makes it crystal clear that $\omega^1 = -\omega^5$. I'll use this notation later in the chapter.

- In the mid-1700s, mathematician Leonhard Euler showed how we can represent roots of unity by their angles around the circle. The number we typically call e is approximately 2.71828. When you take certain powers of e, you get the roots of unity. *Figure 9.13* shows how it works for the eighth roots of unity:

$e^{\left(2\cdot\frac{\pi}{4}\right)}$
$e^{\left(3\cdot\frac{\pi}{2}\right)}$
$e^{\left(1\cdot\frac{\pi}{4}\right)}$
$\frac{\pi}{4}$
$e^{\left(4\cdot\frac{\pi}{4}\right)}$
$e^{\left(0\cdot\frac{\pi}{4}\right)}$
$e^{\left(5\cdot\frac{\pi}{4}\right)}$
$e^{\left(7\cdot\frac{\pi}{4}\right)}$
$e^{\left(6\cdot\frac{\pi}{4}\right)}$

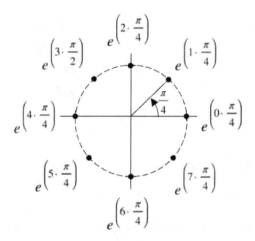

Figure 9.13 – Powers of *e*

Figure 9.13 demonstrates a regular progression from $0 \cdot \frac{\pi}{4}$ to $1 \cdot \frac{\pi}{4}$ to $2 \cdot \frac{\pi}{4}$, and so on. Things don't stop when you reach $7 \cdot \frac{\pi}{4}$. For example, $8 \cdot \frac{\pi}{4}$ is the same as $0 \cdot \frac{\pi}{4}$. (See *Figure 9.14*.)

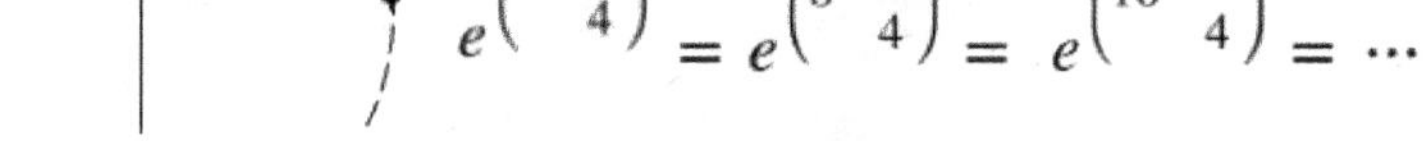

$e^{\left(2\cdot\frac{\pi}{4}\right)} = e^{\left(10\cdot\frac{\pi}{4}\right)} = e^{\left(18\cdot\frac{\pi}{4}\right)} = \cdots$
$e^{\left(1\cdot\frac{\pi}{4}\right)} = e^{\left(9\cdot\frac{\pi}{4}\right)} = e^{\left(17\cdot\frac{\pi}{4}\right)} = \cdots$
$e^{\left(0\cdot\frac{\pi}{4}\right)} = e^{\left(8\cdot\frac{\pi}{4}\right)} = e^{\left(16\cdot\frac{\pi}{4}\right)} = \cdots$

Figure 9.14 – Unending repetition

- If we're not concerned about the progression $0 \cdot \frac{\pi}{4}, 1 \cdot \frac{\pi}{4}, 2 \cdot \frac{\pi}{4}$, we can find simpler labels for the roots of unity. We do this by canceling some terms. After all, $e^{\left(2 \cdot \frac{\pi}{4}\right)}$ is the same as $e^{\frac{\pi}{2}}$. *Figure 9.15* shows you what we get for the fourth roots of unity:

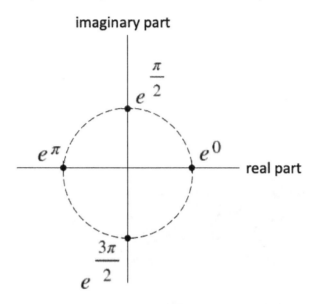

Figure 9.15 – The fourth roots of unity

A funny thing happens when you multiply numbers with exponents. Here's the formula for exponents of e:

$$e^x e^y = e^{(x+y)}$$

To multiply two values, you add their exponents. So, when you multiply the roots of unity, you add their angles, as in the following example:

$$e^{\frac{\pi}{2}} \cdot e^{\pi} = e^{\left(\frac{\pi}{2} + \pi\right)} = e^{\frac{3\pi}{2}}$$

Figure 9.16 shows you how this corresponds to the angles of the circle:

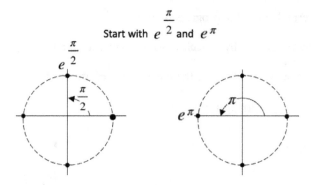

Put the two angles end to end:

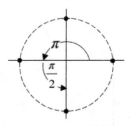

The result is $e^{\frac{\pi}{2}} \cdot e^{\pi}$

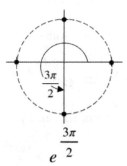

Figure 9.16 – Multiplying roots of unity

Using my circle notation for roots of unity, the multiplication shown in *Figure 9.16* looks like this:

$$\circlearrowright \cdot \ominus = \circledcirc$$

With so many ways to represent roots of unity, we're ready to find the period in a coprime powers sequence. Read on.

Finding a sequence's period

In the section entitled *How Shor's algorithm works*, we saw how knowing the period of a certain repeating sequence of numbers helps an eavesdropper factor a large public key and decrypt a message. How can quantum computing help the eavesdropper discover a sequence's period?

Consider the following sequence of numbers:

> 1, 2, 7, 10, 15, 3, 1, 2, 7, 10, 15, 3, 1, 2, 7, 10, 15, 3, 1, 2, 7, 10, 15, 3

For want of a better name, we'll call this *my sequence*. In my sequence, the **pattern** 1, 2, 7, 10, 15, 3 occurs four times:

- Since the pattern contains six numbers, we say that my sequence's **period** is 6
- Since the pattern occurs four times in my sequence, we say that the pattern's **frequency** is 4

My sequence contains 24 numbers. So, the period and frequency in my sequence are related by the following formula:

$$\text{period} = \frac{24}{\text{frequency}}$$

Mathematicians refer to my sequence's period and its frequency as **dual variables**. If you know the value of one of them, you can derive the value of the other.

In some situations, you know a pattern's frequency, and you use this formula to derive the period. In other situations, you know the period, and you use this formula to obtain the frequency. Shor's algorithm depends on a trick that switches focus between a sequence's period and its frequency.

Imagine a daily diary that's known to contain 2196 words. Each page of the book contains the same six words. (See *Figure 9.17*.)

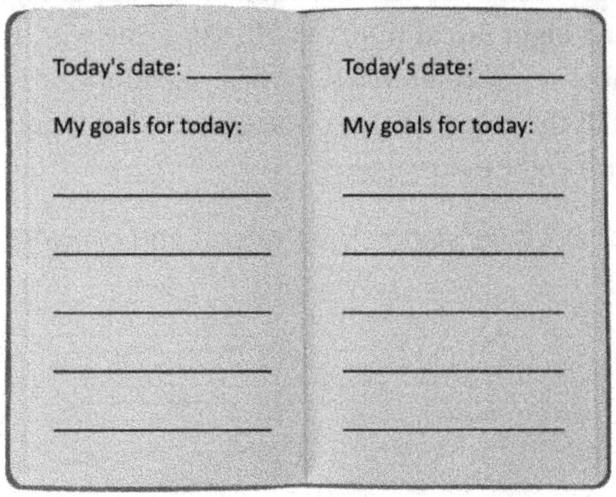

Figure 9.17 – A daily diary (original book photo by engin akyurt on Unsplash)

From this point of view, you have a repeating sequence whose period is 6. With a quick calculation, you determine that the book contains $\frac{2196}{6} = 366$ pages. So, the sequence's frequency is 366.

But what if you don't look at any of the pages? Instead, you turn the book away from you and look at the bottom end of the book. (See *Figure 9.18*.)

Figure 9.18 – Another point of view (photo by Olga Tutunaru on Unsplash)

When you count the sheets in the book's binding, you get 183 sheets. That's 2·183 = 366 pages. Armed with that information, you can deduce that each page has $\frac{2196}{366} = 6$ words. So, the sequence's period is 6.

With Shor's algorithm, a collection of gates performs the equivalent of turning a book sideways.

The QFT matrix

In the section entitled *The connection between complex numbers and circles*, we drew a circle in the plane of complex numbers and marked the circle at equal intervals. In *Figure 9.8*, we had four equal intervals. In *Figure 9.9*, we had eight equal intervals. In the same way, we can have 100 intervals, 1000000 intervals, or any number of equal intervals. In most situations, the number of intervals is very large—larger than the period of the coprime powers sequence that we're analyzing. In this section, we'll illustrate the concepts with eight intervals.

Let the eighth roots of unity be $1, \omega^1, \omega^2, \omega^3, \omega^4, \omega^5, \omega^6, \omega^7$ and consider the following matrix:

$$\frac{1}{\sqrt{8}}\begin{pmatrix} 1 & 1 & 1 & 1 & 1 & 1 & 1 & 1 \\ 1 & \omega^1 & \omega^2 & \omega^3 & \omega^4 & \omega^5 & \omega^6 & \omega^7 \\ 1 & \omega^2 & \omega^4 & \omega^6 & \omega^8 & \omega^{10} & \omega^{12} & \omega^{14} \\ 1 & \omega^3 & \omega^6 & \omega^9 & \omega^{12} & \omega^{15} & \omega^{18} & \omega^{21} \\ 1 & \omega^4 & \omega^8 & \omega^{12} & \omega^{16} & \omega^{20} & \omega^{24} & \omega^{28} \\ 1 & \omega^5 & \omega^{10} & \omega^{15} & \omega^{20} & \omega^{25} & \omega^{30} & \omega^{35} \\ 1 & \omega^6 & \omega^{12} & \omega^{18} & \omega^{24} & \omega^{30} & \omega^{36} & \omega^{42} \\ 1 & \omega^7 & \omega^{14} & \omega^{21} & \omega^{28} & \omega^{35} & \omega^{42} & \omega^{49} \end{pmatrix}$$

This matrix is the 8 × 8 QFT. The matrix has some interesting properties, as described here:

- Each entry's value is $\omega^{(row\ number)(column\ number)}$ with rows and columns numbered from 0 to 7. For example, the entry in row 2, column 3 is $\omega^{2 \cdot 3} = \omega^6$. The entry in row 0, column 4 is $\omega^{0 \cdot 4} = \omega^0 = 1$.

- The entries in row 3 (1, ω^3, ω^6, and so on) are identical to the entries in column 3 (1, ω^3, ω^6, and so on). The same is true for any row and its similarly numbered column.

- The exponents of the entries in each row increase by the row number. For example, the entries in row 5 are ω^0, ω^5, ω^{10}, and so on. The same is true of the entries in each column.

Remember that the roots of unity go around in a circle. So, for example, $\omega^8 = \omega^0 = 1$ and $\omega^9 = \omega^1$. With that in mind, let's rewrite the 8 × 8 QFT matrix with circle notation:

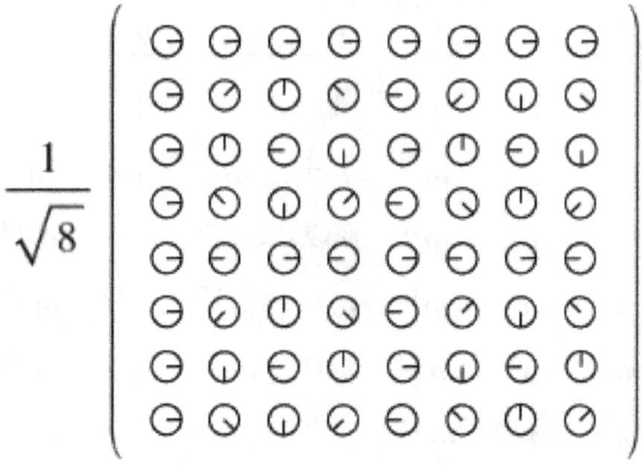

$$\frac{1}{\sqrt{8}} \begin{pmatrix} \circleddash & \circleddash & \circleddash & \circleddash & \circleddash & \circleddash & \circleddash & \circleddash \\ \circleddash & \oslash & \circlearrowleft & \oslash & \circleddash & \oslash & \circlearrowleft & \oslash \\ \circleddash & \circlearrowleft & \circleddash & \circlearrowleft & \circleddash & \circlearrowleft & \circleddash & \circlearrowleft \\ \circleddash & \oslash & \circlearrowleft & \oslash & \circleddash & \oslash & \circlearrowleft & \oslash \\ \circleddash & \circleddash & \circleddash & \circleddash & \circleddash & \circleddash & \circleddash & \circleddash \\ \circleddash & \oslash & \circlearrowleft & \oslash & \circleddash & \oslash & \circlearrowleft & \oslash \\ \circleddash & \circlearrowleft & \circleddash & \circlearrowleft & \circleddash & \circlearrowleft & \circleddash & \circlearrowleft \\ \circleddash & \oslash & \circlearrowleft & \oslash & \circleddash & \oslash & \circlearrowleft & \oslash \end{pmatrix}$$

In this rendering of the QFT matrix, it's easy to see which values are negatives of one another. For example, $\circleddash$ represents the value 1, and $\ominus$ represents the value −1. So, $\ominus = -\circleddash$. In the same way, $\circlearrowleft = -\circlearrowleft$ and $\oslash = -\oslash$. Any two circles that point in opposite directions are negatives of one another. With this observation, we can see that some products may cancel one another out. *Figure 9.19* illustrates this idea:

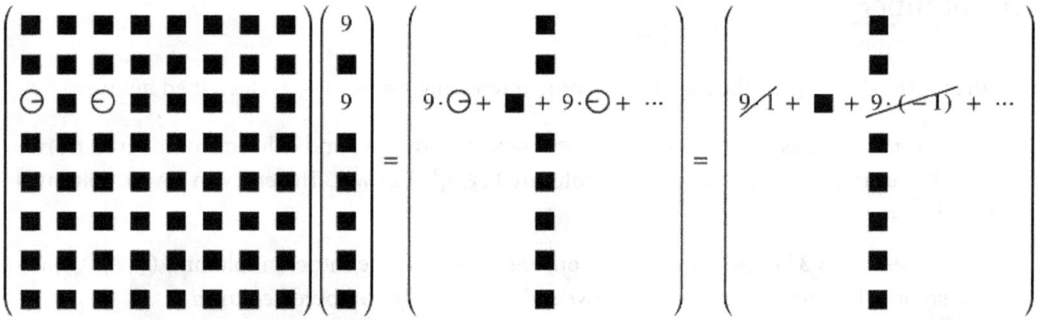

Figure 9.19 – Cancellation of values with the QFT

In *Figure 9.19*, you can ignore all the entries that contain little dark squares. With the entries that aren't dark squares, two opposite circles in the QFT matrix match up with two equal values (two 9s) in the vector. So, the two terms cancel one another.

> **Important note**
>
> Notice how we omit the $\frac{1}{\sqrt{8}}$ that belongs at the start of *Figure 9.19*. For the concepts in this section, the $\frac{1}{\sqrt{8}}$ (and other such scalar multipliers) don't matter.

Figure 9.20 has another example:

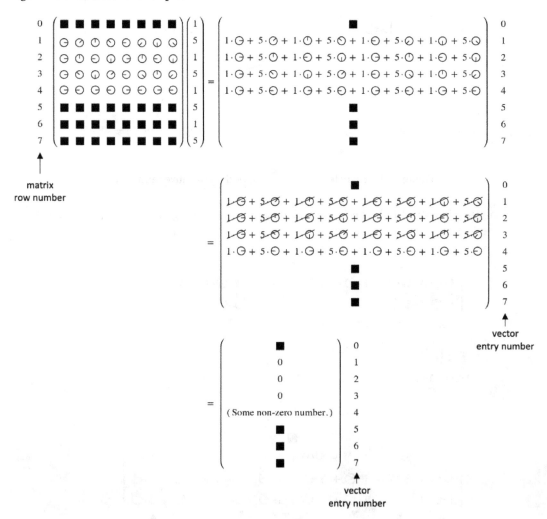

Figure 9.20 – Some rows' values cancel; others don't

In *Figure 9.20*, we multiply the QFT matrix by a vector with a repeating 1, 5, 1, 5, ... pattern. Since the pattern is 1, 5, the sequence's period is 2.

Figure 9.20 illustrates an interesting phenomenon about the vector we obtain when we apply the QFT. In the summation for entry number 1, every term cancels with some other term. (See *Figure 9.21*.)

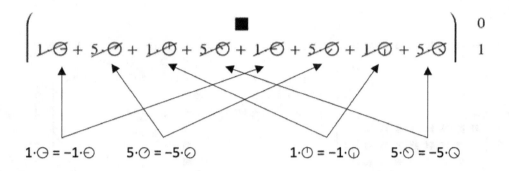

Figure 9.21 – Every term has a corresponding negative term

So, the final value in entry 1 is 0. This is not accidental. The same thing happens for entries 2 and 3. (See *Figure 9.22*.)

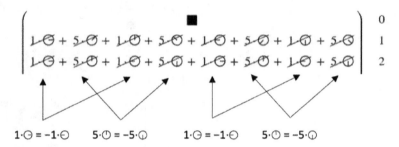

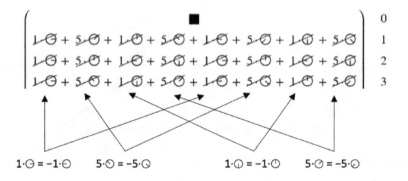

Figure 9.22 – More cancellation

So, in *Figure 9.20*, the final values of entries 2 and 3 are also 0. But no such thing happens with entry 4. (See *Figure 9.23*.)

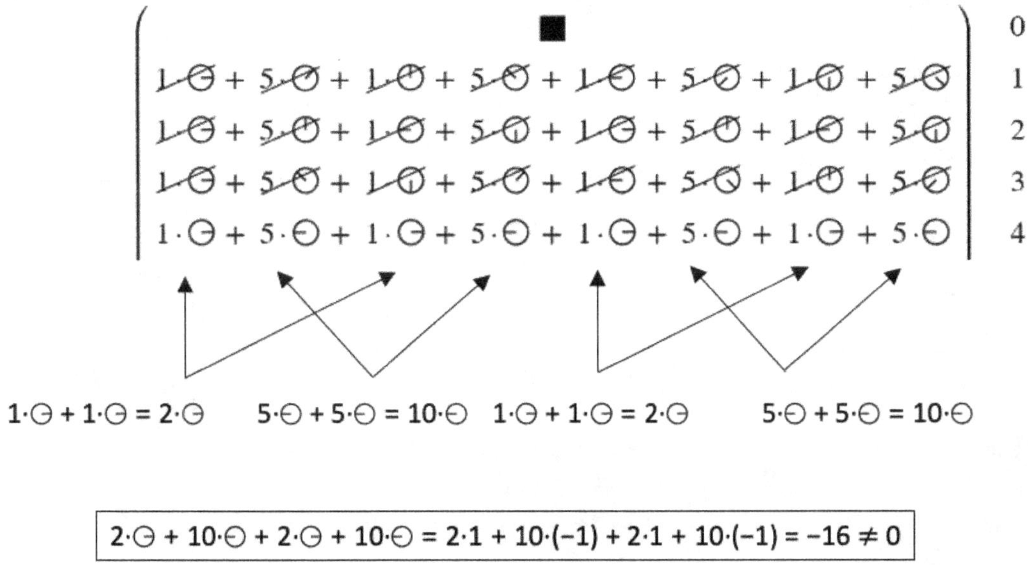

$$\left(\begin{array}{c} \blacksquare \\ 1{\cdot}\ominus + 5{\cdot}\ominus + 1{\cdot}\ominus + 5{\cdot}\ominus + 1{\cdot}\ominus + 5{\cdot}\ominus + 1{\cdot}\ominus + 5{\cdot}\ominus \\ 1{\cdot}\ominus + 5{\cdot}\ominus + 1{\cdot}\ominus + 5{\cdot}\ominus + 1{\cdot}\ominus + 5{\cdot}\ominus + 1{\cdot}\ominus + 5{\cdot}\ominus \\ 1{\cdot}\ominus + 5{\cdot}\ominus + 1{\cdot}\ominus + 5{\cdot}\ominus + 1{\cdot}\ominus + 5{\cdot}\ominus + 1{\cdot}\ominus + 5{\cdot}\ominus \\ 1{\cdot}\ominus + 5{\cdot}\ominus + 1{\cdot}\ominus + 5{\cdot}\ominus + 1{\cdot}\ominus + 5{\cdot}\ominus + 1{\cdot}\ominus + 5{\cdot}\ominus \end{array}\right) \begin{array}{c} 0 \\ 1 \\ 2 \\ 3 \\ 4 \end{array}$$

$1{\cdot}\ominus + 1{\cdot}\ominus = 2{\cdot}\ominus$ $5{\cdot}\ominus + 5{\cdot}\ominus = 10{\cdot}\ominus$ $1{\cdot}\ominus + 1{\cdot}\ominus = 2{\cdot}\ominus$ $5{\cdot}\ominus + 5{\cdot}\ominus = 10{\cdot}\ominus$

$2{\cdot}\ominus + 10{\cdot}\ominus + 2{\cdot}\ominus + 10{\cdot}\ominus = 2{\cdot}1 + 10{\cdot}(-1) + 2{\cdot}1 + 10{\cdot}(-1) = -16 \neq 0$

Figure 9.23 – No cancellation in row 4

When you apply the QFT matrix to a vector with a repeating sequence, almost all the resulting entries have values that cancel out to 0. The only exceptions are entries such as entry number 4 in *Figure 9.23*. With entry 4, the repeating pattern of the numbers 1, 5, 1, 5, ... aligns with the repeating pattern of the circles $\ominus, \ominus, \ominus, \ominus, \ldots$.

Figure 9.24 has one more example:

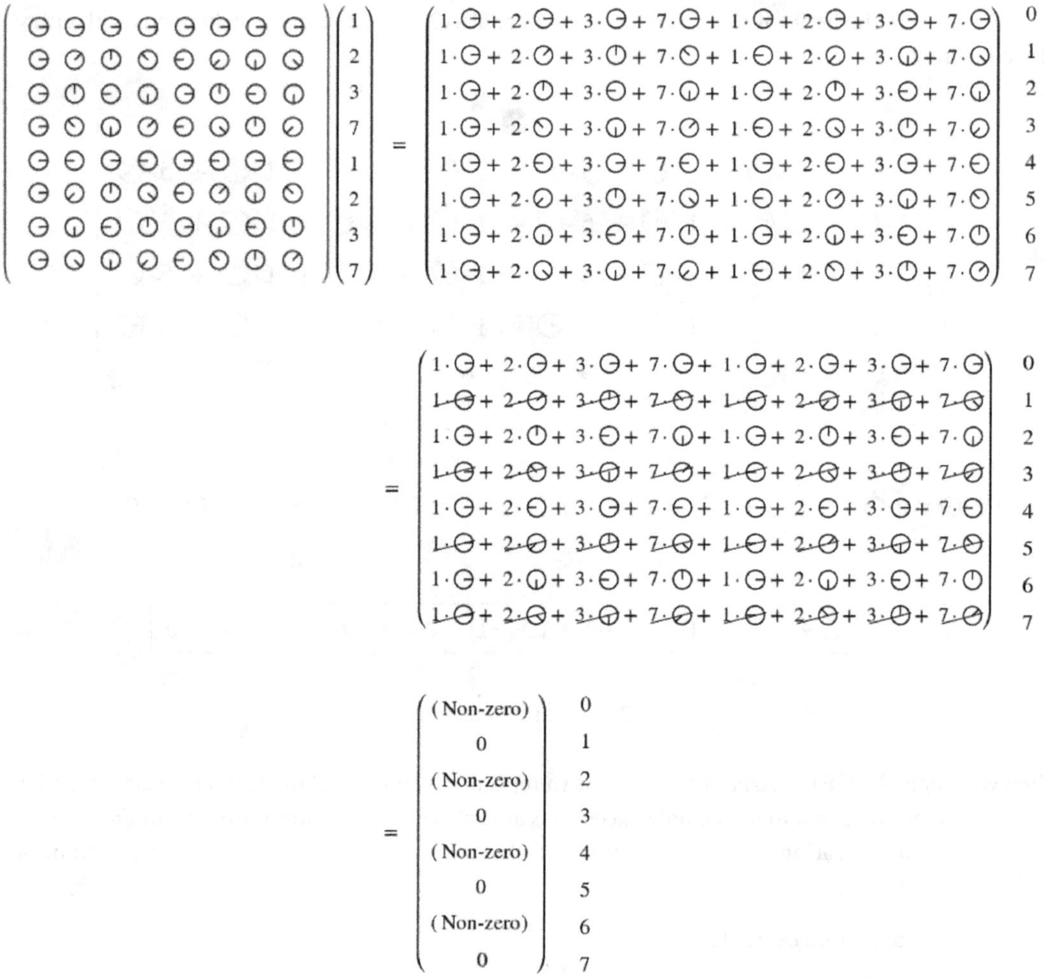

$$
\begin{pmatrix}
\ominus & \ominus & \ominus & \ominus & \ominus & \ominus & \ominus & \ominus \\
\ominus & \oslash & \oslash & \oslash & \oslash & \oslash & \oslash & \oslash \\
\ominus & \oslash & \ominus & \oslash & \ominus & \oslash & \ominus & \oslash \\
\ominus & \oslash & \oslash & \oslash & \ominus & \oslash & \oslash & \oslash \\
\ominus & \ominus & \ominus & \ominus & \ominus & \ominus & \ominus & \ominus \\
\ominus & \oslash & \oslash & \oslash & \ominus & \oslash & \oslash & \oslash \\
\ominus & \oslash & \ominus & \oslash & \ominus & \oslash & \ominus & \oslash \\
\ominus & \oslash & \oslash & \oslash & \ominus & \oslash & \oslash & \oslash
\end{pmatrix}
\begin{pmatrix} 1 \\ 2 \\ 3 \\ 7 \\ 1 \\ 2 \\ 3 \\ 7 \end{pmatrix}
=
\begin{pmatrix}
1\cdot\ominus + 2\cdot\ominus + 3\cdot\ominus + 7\cdot\ominus + 1\cdot\ominus + 2\cdot\ominus + 3\cdot\ominus + 7\cdot\ominus \\
1\cdot\ominus + 2\cdot\oslash + 3\cdot\oslash + 7\cdot\oslash + 1\cdot\ominus + 2\cdot\oslash + 3\cdot\oslash + 7\cdot\oslash \\
1\cdot\ominus + 2\cdot\oslash + 3\cdot\ominus + 7\cdot\oslash + 1\cdot\ominus + 2\cdot\oslash + 3\cdot\ominus + 7\cdot\oslash \\
1\cdot\ominus + 2\cdot\oslash + 3\cdot\oslash + 7\cdot\oslash + 1\cdot\ominus + 2\cdot\oslash + 3\cdot\oslash + 7\cdot\oslash \\
1\cdot\ominus + 2\cdot\ominus + 3\cdot\ominus + 7\cdot\ominus + 1\cdot\ominus + 2\cdot\ominus + 3\cdot\ominus + 7\cdot\ominus \\
1\cdot\ominus + 2\cdot\oslash + 3\cdot\oslash + 7\cdot\oslash + 1\cdot\ominus + 2\cdot\oslash + 3\cdot\oslash + 7\cdot\oslash \\
1\cdot\ominus + 2\cdot\oslash + 3\cdot\ominus + 7\cdot\oslash + 1\cdot\ominus + 2\cdot\oslash + 3\cdot\ominus + 7\cdot\oslash \\
1\cdot\ominus + 2\cdot\oslash + 3\cdot\oslash + 7\cdot\oslash + 1\cdot\ominus + 2\cdot\oslash + 3\cdot\oslash + 7\cdot\oslash
\end{pmatrix}
\begin{matrix} 0 \\ 1 \\ 2 \\ 3 \\ 4 \\ 5 \\ 6 \\ 7 \end{matrix}
$$

$$
=
\begin{pmatrix}
1\cdot\ominus + 2\cdot\ominus + 3\cdot\ominus + 7\cdot\ominus + 1\cdot\ominus + 2\cdot\ominus + 3\cdot\ominus + 7\cdot\ominus \\
1\cdot\ominus + 2\cdot\ominus + 3\cdot\ominus + 7\cdot\ominus + 1\cdot\ominus + 2\cdot\ominus + 3\cdot\ominus + 7\cdot\ominus \\
1\cdot\ominus + 2\cdot\oslash + 3\cdot\ominus + 7\cdot\oslash + 1\cdot\ominus + 2\cdot\oslash + 3\cdot\ominus + 7\cdot\oslash \\
1\cdot\ominus + 2\cdot\ominus + 3\cdot\ominus + 7\cdot\ominus + 1\cdot\ominus + 2\cdot\ominus + 3\cdot\ominus + 7\cdot\ominus \\
1\cdot\ominus + 2\cdot\ominus + 3\cdot\ominus + 7\cdot\ominus + 1\cdot\ominus + 2\cdot\ominus + 3\cdot\ominus + 7\cdot\ominus \\
1\cdot\ominus + 2\cdot\ominus + 3\cdot\ominus + 7\cdot\ominus + 1\cdot\ominus + 2\cdot\ominus + 3\cdot\ominus + 7\cdot\ominus \\
1\cdot\ominus + 2\cdot\oslash + 3\cdot\ominus + 7\cdot\oslash + 1\cdot\ominus + 2\cdot\oslash + 3\cdot\ominus + 7\cdot\oslash \\
1\cdot\ominus + 2\cdot\ominus + 3\cdot\ominus + 7\cdot\ominus + 1\cdot\ominus + 2\cdot\ominus + 3\cdot\ominus + 7\cdot\ominus
\end{pmatrix}
\begin{matrix} 0 \\ 1 \\ 2 \\ 3 \\ 4 \\ 5 \\ 6 \\ 7 \end{matrix}
$$

$$
=
\begin{pmatrix}
(\text{Non-zero}) \\
0 \\
(\text{Non-zero}) \\
0 \\
(\text{Non-zero}) \\
0 \\
(\text{Non-zero}) \\
0
\end{pmatrix}
\begin{matrix} 0 \\ 1 \\ 2 \\ 3 \\ 4 \\ 5 \\ 6 \\ 7 \end{matrix}
$$

Figure 9.24 – Zero and nonzero vector values

The result in *Figure 9.24* has four nonzero entries where the following applies:

- In entry 2, the repeating pattern of the numbers 1, 2, 3, 7 aligns with the repeating pattern of circles ⊖, ⊘, ⊖, ⊘.

- In entry 6, the repeating pattern of the numbers 1, 2, 3, 7 aligns with a slightly different repeating pattern of circles ⊖, ⊘, ⊖, ⊘.

- In entry 4, the repeating pattern of the numbers 1, 2, 3, 7 aligns with the repeating pattern of circles ⊖, ⊖, ⊖, ⊖. Within this pattern of four circles is a smaller pattern ⊖, ⊖ of two circles, but that smaller pattern doesn't affect the four-number alignment.

- In entry 1, the repeating pattern of the numbers 1, 2, 3, 7 aligns with the repeating pattern of circles ⊝, ⊝, ⊝, ⊝. Within this pattern of four circles is a smaller pattern of one circle ⊝, but that smaller pattern doesn't affect the four-number alignment.

And now, we observe something that forms the core of Shor's algorithm:

- At the top of *Figure 9.20*, the vector sequence 1, 5, 1, 5, 1, 5, 1, 5 has frequency 4. At the bottom of the figure, ignore entry 0. Then, the smallest numbered entry to have a nonzero result is entry 4.

- At the top of *Figure 9.24*, the vector sequence 1, 2, 3, 7, 1, 2, 3, 7 has frequency 2. At the bottom of the figure, ignore entry 0. Then, the smallest numbered entry to have a nonzero result is entry 2.

In both cases, the smallest nonzero entry tells us the frequency in the vector sequence. So, here's the sum and substance of Shor's algorithm:

Using the tricks in the *Repeated squaring* section, Eve can generate a coprime powers sequence from Alice's public key. With a QFT matrix of a known size, she can find the sequence's frequency. By dividing that frequency into the width of the matrix, Eve finds the sequence's period. Then, using the techniques in the section entitled *The role of a period in factoring a number*, Eve discovers the private key. She uses the private key to decrypt Bob's message. Oy, vey!

At the start of this *Finding a sequence's period* section, we described the act of observing a book from two points of view. When you stare directly at a page, you can count words to find the book's period. When you turn the book sideways, you can count pages to find the book's frequency. Turning the book sideways is like applying the QFT.

One problem with this book analogy is imagining how difficult it would be to count sheets. With 183 sheets stacked tightly against one another, you couldn't just glance at the edge of the book and call out the number of pages. But quantum computing doesn't work that way. In addition to changing the point of view, the QFT filters out all but the most salient information.

I've mocked up some histograms in *Figure 9.25*. With the histogram on the left, you have to count upward from 00 to find the start of the repeating pattern. But with the histogram on the right, all the values between 000 and 183 are gone. Finding the frequency is a one-step process:

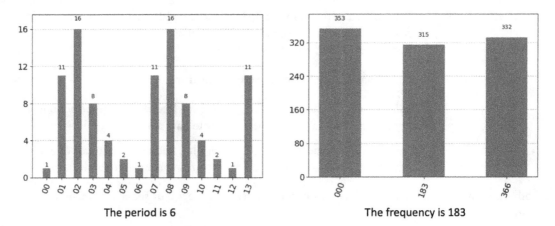

The period is 6   The frequency is 183

Figure 9.25 – Mocked-up histograms for the diary analogy

As with many things in this chapter, the point that I make in *Figure 9.25* is greatly oversimplified. Shor's period-finding technique is an intricate process with many subtle twists and turns. The next section touches on a few of those details.

Shoring up your knowledge

Understanding Shor's algorithm can be difficult because examples with manageable-size numbers are hard to find. For instance, a minimal circuit that factors 15 with 11 as its coprime may involve five qubits. Wielding five qubits at once means multiplying 32×32 matrices by one another. Each matrix contains 1024 complex numbers. That's too many numbers for one example in a book.

You can overcome the conceptual difficulties using summations and linear algebra, and I encourage you to study more about these approaches. In the meantime, this section describes some aspects of Shor's algorithm that previous sections glossed over.

At some future date, when we have quantum computers that can crack real RSA encryption problems, those computers will probably have thousands of qubits. Alice will start with a public key, n, that has 2048 bits. To attack that key with Shor's algorithm, Eve's circuit will implement an $N \times N$ matrix, where N is a power of 2 between n^2 and $2n^2$.

For any value N, the $N \times N$ QFT is this:

$$\frac{1}{\sqrt{N}} \begin{pmatrix} 1 & 1 & 1 & 1 & 1 & 1 \\ 1 & \omega & \omega^2 & \omega^3 & \cdots & \omega^{N-1} \\ 1 & \omega^2 & \omega^4 & \omega^6 & \cdots & \omega^{2(N-1)} \\ 1 & \omega^3 & \omega^6 & \omega^9 & \cdots & \omega^{3(N-1)} \\ \vdots & \vdots & \vdots & \vdots & \ddots & \vdots \\ 1 & \omega^{N-1} & \omega^{2(N-1)} & \omega^{3(N-1)} & \cdots & \omega^{(N-1)(N-1)} \end{pmatrix}$$

In this matrix, ω stands for the Nth root of unity. As in the section entitled *The QFT matrix*, the exponent of each ω entry is the entry's column number times its row number. Each row goes around in circles at one pace or another, and we multiply these circular values by a vector's values. Roughly speaking, that vector contains a coprime powers sequence of the kind we generated in the *How Shor's algorithm works* section.

To be more precise, the vector contains a sequence whose values vary *in the same proportions* as the values in our coprime powers sequence. For example, imagine that our coprime powers sequence contains the numbers 1, 5, 1, 5, 1, 5, 1, 5. To represent this information with qubit amplitudes, we must divide each number so that the sum of squares adds up to 1. Since the sequence's squares add up to 104, we divide each number by $\sqrt{104} \approx 10.2$. So, the vector that represents the state of three qubits (the vector we multiply by the QFT) is this:

$$\begin{pmatrix} 1/\sqrt{104} \\ 5/\sqrt{104} \\ 1/\sqrt{104} \\ 5/\sqrt{104} \\ 1/\sqrt{104} \\ 5/\sqrt{104} \\ 1/\sqrt{104} \\ 5/\sqrt{104} \end{pmatrix} = \frac{1}{\sqrt{104}} \begin{pmatrix} 1 \\ 5 \\ 1 \\ 5 \\ 1 \\ 5 \\ 1 \\ 5 \end{pmatrix}$$

We put three qubits into this state, and send the qubits through circuitry that implements the QFT matrix.

After representing the coprime power sequence as a state vector for qubits, we still have no guarantee that the pattern in the vector occurs an integer number of times. What if a 256×256 matrix meets a state vector with period 6? (This reads like the lead-in to a corny joke!) The vector must have 256 entries, so the pattern will occur $\frac{256}{6} = 42\frac{2}{3}$ times. But after applying the QFT, we can't end up with a vector whose $42\frac{2}{3}$th entry is nonzero. There's no such thing as a $42\frac{2}{3}$th entry. In his original paper, Shor deals with this problem.

Here's another potential problem. In the *How Shor's algorithm works* section, you find a sequence's period and calculate *coprime*$^{period/2}$. What if the period is an odd number? The trick that we describe in the section entitled *The role of a period in factoring a number* doesn't work if the exponent *period*/2 is a fraction.

In this case, you throw away everything you've done so far and start again with a different coprime value. The time you've wasted turns out to be insignificant, and the likelihood that the next coprime yields an even-number period is pretty good.

With that said, it's time for some code.

Illustrating Shor's algorithm with Qiskit code

In this section, we present two Qiskit programs. One program illustrates, in a straightforward manner, the application of Qiskit's QFT function. The other uses special tricks to construct a circuit that scales for large numbers. Neither program implements Shor's algorithm in complete detail.

Testing the QFT

We begin this example with the required `import` statements:

```
from qiskit import QuantumCircuit, Aer, execute
from qiskit.circuit.library import QFT
from qiskit.tools.visualization import plot_histogram
import numpy as np
```

With public key 15 and coprime 7, we create a `vector` containing eight values of 7^n % 15:

```
public_key = 15
coprime = 7
#coprime = 11

vector = []
for i in range(8):
    vector.append(coprime**i % public_key)
```

```
norm = np.linalg.norm(vector)
statevector = vector / norm

print('vector:')
print(vector)
print()
print('statevector:')
print(statevector)
```

The output of this code is shown in *Figure 9.26*:

```
vector:
[1, 7, 4, 13, 1, 7, 4, 13]

statevector:
[0.04612656 0.32288592 0.18450624 0.59964529 0.04612656 0.32288592
 0.18450624 0.59964529]
```

Figure 9.26 – Vectors for a coprime powers sequence

The values in the `vector` are 1, 7, 4, and so on. The sum of these numbers' squares adds up to much more than 1. So, we have to divide all the numbers by the sum of their squares. We use NumPy's `linalg.norm` function to find that sum. After doing the division, the values in the new `statevector` have the same proportions as those in the original `vector`, and the sum of the squares in the `statevector` is 1. This allows us to initialize three qubits in a way that's described by the `statevector`. Here's the code:

```
circ = QuantumCircuit(3)
circ.initialize(statevector)
```

Our next step is to send these qubits through a QFT. Qiskit has its own QFT class, so we can create an instance of that class and append it to the existing circuit, like so:

```
circ.append(QFT(3), [0, 1, 2])
circ.measure_all()
display(circ.draw('latex'))
```

The resulting output is shown in *Figure 9.27*:

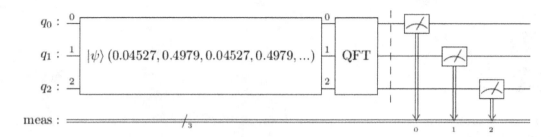

Figure 9.27 – A circuit to test the QFT

To run the circuit, we use some familiar Python code:

```
device = Aer.get_backend('qasm_simulator')
job = execute(circ,backend = device,shots = 1000)
print(job.job_id())
result = job.result()
counts = result.get_counts(circ)
print(counts)
display(plot_histogram(counts))
```

The output of the run is a chart. It looks like the one shown in *Figure 9.28*:

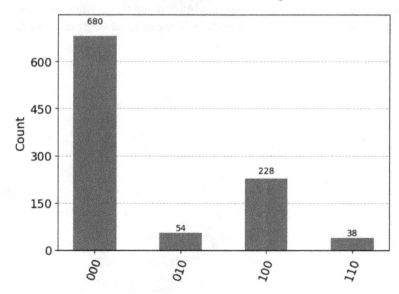

Figure 9.28 – QFT shows a frequency of 2

If we turn the histogram in *Figure 9.28* into a vector, we get the following:

$$\frac{1}{1000} \begin{pmatrix} 680 \\ 0 \\ 54 \\ 0 \\ 288 \\ 0 \\ 110 \\ 0 \end{pmatrix}$$

Ignoring entry 0 (as usual), the first nonzero entry is entry 2. And, if we look at the frequency of this problem's coprime powers sequence, we see the same number 2. So, the results in *Figure 9.28* are consistent with what we know about the QFT.

If we start the example with `public_key=15` and `coprime=11`, we get the vector values and histogram shown in *Figure 9.29*:

```
vector:
[1, 11, 1, 11, 1, 11, 1, 11]

statevector:
[0.04526787 0.4979466  0.04526787 0.4979466  0.04526787 0.4979466
 0.04526787 0.4979466 ]
```

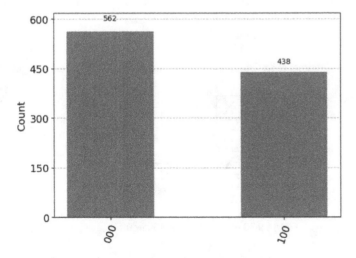

Figure 9.29 – QFT shows a frequency of 4

Once again, our suspicions about the behavior of the QFT are borne out. The frequency of the coprime powers sequence is 4, and the smallest measurement of qubits (other than |000⟩) is the binary representation for 4. For Eve, all is well. (The rest of us are in big trouble.)

Another implementation of Shor's algorithm

If you search for implementations of Shor's algorithm, you'll encounter a technique called **Quantum Phase Estimation (QPE)** [3]. These implementations are different from the one in the previous section's code. As with the previous section's example, QPE uses coprime powers sequences, period finding, QFT matrices, and all the other concepts that we've covered in this chapter. But QPE takes advantage of superposition and phase kickback to represent the coprime powers sequence more efficiently. In this section, we implement Shor's algorithm with QPE and give the *briefest* of explanations to describe how it works. (For another example of a QPE circuit, see *Question 10* at the end of this chapter.)

We start with the usual list of `import` statements:

```
from qiskit import QuantumCircuit, Aer, execute
from qiskit.circuit.library import QFT
from qiskit.visualization import plot_histogram
```

Next, we create a small circuit:

```
_7k_mod15 = QuantumCircuit(4)
_7k_mod15.x([0, 1, 2, 3])
_7k_mod15.swap(1, 2)
_7k_mod15.swap(2, 3)
_7k_mod15.swap(0, 3)
display(_7k_mod15.draw('latex'))
```

This circuit has three SWAP gates. (See *Chapter 4.*) When we display the circuit, we get the drawing shown in *Figure 9.30*:

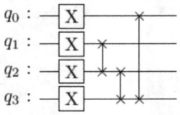

Figure 9.30 – The _7k_mod15 circuit

What in the world does this circuit do? To find out, let's input the number 6 in its binary form 0110. The circuit's four X gates reverse each of the bits, giving us 1001. Then, the swaps do what's called a **right circular shift**. (See *Figure 9.31*.)

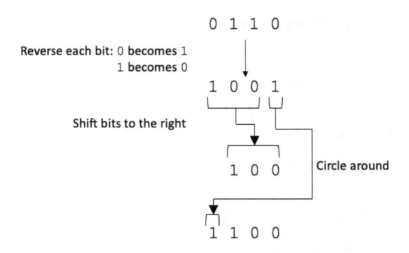

Figure 9.31 – Right circular shift

The result is the binary representation of the number 12. But 12 is the same as 7·6 % 15.

Try it with some other inputs besides the number 6. For any input number, *k*, this circuit calculates 7*k* % 15. That's why I've named the circuit _7k_mod15.

Here's a function that applies the _7k_mod15 function n times:

```
def _7EPXn_mod15(n):
    circ = QuantumCircuit(4)
    for k in range(n):
        circ = circ.compose(_7k_mod15, qubits=[0, 1, 2, 3])

    gate = circ.to_gate(label='(7^' + str(n) + ') mod 15')
    return gate.control(1, ctrl_state='1')
```

The _7EPXn_mod15 function creates a quantum gate in which _7k_mod15 is applied n times. This is useful for finding powers of 7. (See *Figure 9.32*.)

$$1 \xrightarrow{\text{times } 7} 7 \xrightarrow{\text{times } 7} 7^2 \xrightarrow{\text{times } 7} 7^3 \xrightarrow{\text{times } 7} \cdots$$

Figure 9.32 – Powers of 7

As its final act, the _7EPXn_mod15 function applies Qiskit's control method. We introduced the control method in *Chapter 8*. This method does for the _7EPXn_mod15 gate what CNOT does for the X gate. The control method enables another qubit's value to determine whether the _7EPXn_mod15 gate will be activated.

In the next cell of the notebook, we put the pieces together:

```
circ = QuantumCircuit(7, 3)
circ.h([0, 1, 2])
circ.x(3)
circ.barrier()

circ.append(_7EPXn_mod15(1), [0, 3, 4, 5, 6])
circ.append(_7EPXn_mod15(2), [1, 3, 4, 5, 6])
circ.append(_7EPXn_mod15(4), [2, 3, 4, 5, 6]) # NOT NEEDED

circ.append(QFT(3).inverse(), [0, 1, 2])
circ.measure([0, 1, 2], [0, 1, 2])

display(circ.draw('latex'))
```

Figure 9.33 contains an annotated version of this code's output:

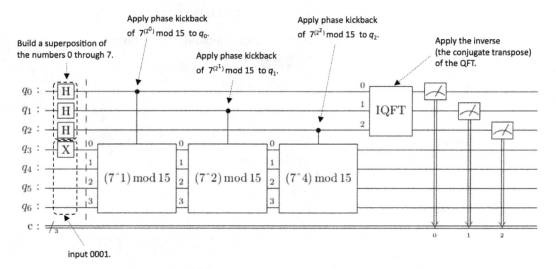

Figure 9.33 – Shor's algorithm with QPE

Let's start by considering the three Hadamard gates on the left side of *Figure 9.33*. After passing through these H gates, the top three qubits are in a superposition of the $|000\rangle$, $|001\rangle$, $|010\rangle$, $|011\rangle$, $|100\rangle$, $|101\rangle$, $|110\rangle$, and $|111\rangle$ states. These Hadamard gates ensure that all coprime powers, 2^0 through 2^7, will be represented in the calculation.

Also on the left, an X gate puts the bottom four qubits into the state $|0001\rangle$. That starts the ball rolling to obtain 7^1 mod 15, which feeds into 7^2 mod 15, which finally feeds into 7^4 mod 15. Each of the control qubits (q_0, q_1, and q_2) feels the effect of its respective target gate. (Strictly speaking, the 7^4 mod 15 gate isn't needed, because 7^4 mod 15 = 1.)

Near the right side of *Figure 9.33*, we use Qiskit's QFT class. We create an instance of the class and apply the class's inverse method to get the inverse of the QFT matrix (labeled **IQFT**). The inverse of the QFT matrix has the same frequency-finding properties as the QFT, but the complex conjugate is easier to implement with quantum gates.

We can run this circuit using Python code from the end of the Testing the QFT section. When we do, we get the kind of output shown in *Figure 9.34*.

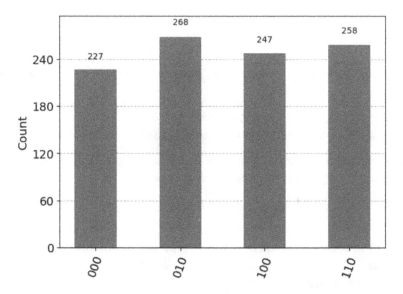

Figure 9.34 – The output of the circuit shown in Figure 9.33

No doubt about it. Eve is on the verge of breaking the RSA encryption scheme. We better get to work devising Shor-resistant cryptography.

Summary

Shor's algorithm is the crowning glory in the theory of gate-based quantum computing. Using Shor's algorithm, a quantum computer can do what no classical computer could ever have time to do. The algorithm quickly finds the period of a coprime powers sequence and uses information about that period to factor unimaginably large numbers. That's the good news.

The bad news is that the most obvious use of Shor's algorithm is for unsavory purposes. Many applications in use today include some form of RSA encryption. And with Shor's algorithm, a hacker can easily break the RSA encryption scheme. For this reason, the **National Institute of Standards and Technology (NIST)** in the United States has been running an initiative to develop quantum-proof post-quantum cryptography schemes [4]. Someday, in the not-too-distant future, the battle over your security and privacy will be fought with quantum computers and even more robust cryptographic algorithms.

This book focuses on circuits and gates. That's one way to do quantum computing, but it's not the only way. The book's final chapter describes a few other ways to leverage the bizarre features of quantum physics.

Further reading

[1] *P. W. Shor. Algorithms for quantum computation: discrete logarithms and factoring. Proceedings 35th Annual Symposium on Foundations of Computer Science, Santa Fe, NM, USA, 1994, pp. 124-134, doi: 10.1109/SFCS.1994.365700.*

[2] *Rivest, R. L., Shamir, A.,* and *Adleman, L. M.* (*1977, August*). *A method for obtaining digital signatures and public-key cryptosystems*: `https://web.williams.edu/Mathematics/lg5/302/RSA.pdf`

[3] **QPE**: `https://github.com/qiskit-community/qiskit-textbook/blob/main/content/ch-algorithms/quantum-phase-estimation.ipynb`

[4] *Post-Quantum Cryptography | Round 4 Submissions*: `https://csrc.nist.gov/projects/post-quantum-cryptography/round-4-submissions`

Questions

1. Using a pencil and paper or a hand calculator, find the factors of 14 using 3 as a coprime. (Use only the technique shown in the section entitled *The role of a period in factoring a number*. Don't try applying the QFT.)

2. Using a pencil and paper or a hand calculator, find the factors of 35 using 2 as a coprime. (Again, use only the technique shown in the section entitled *The role of a period in factoring a number*.)

3. Write the value of $e^{\left(71 \cdot \frac{\pi}{4}\right)}$ using $a + bi$ notation.

4. Write down the entries of a 4×4 QFT matrix. Use the notation for complex numbers that's shown in *Figure 9.8*. Then, use the 4×4 QFT matrix to find the entries in the QFT† matrix.

5. Write down the entries of a 2×2 QFT matrix. Does this matrix look familiar?

6. Fill in the missing values in *Figure 9.20*.

7. Why isn't 3 useful as a coprime when you try to factor 22?

8. Starting with the powers of 7 shown in the *How Shor's algorithm works* section, use repeated squaring to find 7^{11} % 15.

9. Alice multiplies the prime numbers 11 and 13 to obtain a public key. For her RSA coprime, she selects 7. Use the public key and the coprime to encrypt the letter a. (The ASCII code for a is 97.)

Find Alice's private key using a classical computer program, a generative AI bot, or (if you're very ambitious) brute force.

10. Run the following code several times with different values for the `phase` variable:

```
phase = None
circ = QuantumCircuit(2, 1)
circ.h([0, 1])
circ.rz(phase, 1)
circ.cnot(0, 1)
circ.h(0)
circ.measure([0], [0])
display(circ.draw('latex'))

provider = IBMProvider()
device = provider.get_backend('ibmq_qasm_simulator')
job = execute(circ, device, shots=1000)
result = job.result()
counts = result.get_counts(circ)
print(counts)
```

Use the results of these runs to verify the validity of the following formula:

$$\text{phase} \approx \arccos\left(2\left(\frac{\text{count\_of\_zeros\_measured}}{\text{number\_of\_shots}}\right) - 1\right)$$

$$\text{phase} \approx \arccos\left(2\left(\frac{\text{count\_of\_zeros\_measured}}{\text{number\_of\_shots}}\right) - 1\right)$$

In this code, the `rz` gate rotates the bottom qubit around the Bloch sphere's equator. This formula tells us that we can estimate the angle of rotation (the bottom qubit's phase) by measuring the top qubit. This is an extremely simple example of QPE.

Part 4
Beyond Gate-Based Quantum Computing

The focus of this book is on only one kind of quantum computer: the gate-based model. In this part, we will introduce four other models. We will describe the way each model's design differs from the gate-based model's design. We will also describe the kinds of problems each model can solve.

This part has the following chapter:

- *Chapter 10, Some Other Directions for Quantum Computing*

10
Some Other Directions for Quantum Computing

This book is about gate-based quantum computing. In each example, we prepare qubits and apply operations with gates. Most of the gates perform reversible quantum operations. The main exceptions to the reversibility rule are the measurement gates, which destroy superposition and yield solutions to our intricate, mathematical problems.

Gate-based quantum computing is great for certain problems such as cracking RSA encryption, but, as far as we know, many other problems are more easily solved using classical computing. To search for a value in an unsorted list, you can run Grover's algorithm on a quantum computer, but to find out how much tax you owe the government this year, you should run a Java program on a classical computer. (In fact, doing all your tax calculations by hand would probably be more efficient than relying on any kind of quantum computer for help.)

Nevertheless, gate-based quantum computing is an example of **general-purpose computing** because the gate-based model isn't designed with any particular kind of application in mind. When you work with quantum gates, you think about the logic behind the problem at hand. And logic applies to any kind of problem, whether it's RSA encryption, tax calculation, or sending robots shaped like Barbie™ dolls to an outpost on the moon.

In contrast, a **special-purpose computer** is one whose design targets a specific kind of problem. Think about the **Graphics Processing Unit** (**GPU**) inside your own computer. A GPU's circuitry is optimized for graphics rendering and other high-performance tasks, but you can't compile a Python program on a GPU (or, at least, you shouldn't). A GPU is simply a helper for a CPU or other general-purpose hardware.

In the same way, companies have designed and built some special-purpose quantum computers. With many of these computers, you don't start by defining a circuit and its gates. Instead, you create some other structure whose properties take advantage of superposition and entanglement. This chapter offers a glimpse at a few of these special-purpose quantum projects.

We'll cover the following topics in this chapter:

- What reducibility is

- Quantum simulation

- Quantum annealing

- Quantum neural nets

- Solving unsolvable problems

What is reducibility?

In *The Karate Kid* (Sony Pictures, 1984), Mr. Miyagi teaches Daniel to do defensive blocks by having him repeatedly wax Miyagi's car. The idea is to learn how to block an attack, starting by training yourself to apply and remove car wax. Start with the problem you want to solve (blocking an attacker's approach) and turn it into a different problem (rubbing metal to make it shine). At first sight, the two problems seem to be unrelated, but in the movie, Daniel learns that the skills underlying both problems are identical. A solution to the waxing problem tells Daniel how to solve the defensive blocking problem.

In the formal terminology of algorithms, we might say that the defensive blocking problem is **reducible** to the car-waxing problem. The word *reducible* suggests that, in some way, waxing a car is simpler than defensive blocking. Indeed, the simplicity of polishing metal is what helps Daniel to learn movements so naturally.

In the same way, we can take business and scientific problems and turn them into simpler problems. For example, in microeconomics the formula for Total Revenue is:

$$Total\ Revenue\ =\ Price\ \cdot\ Quantity\ in\ Demand$$

Compare this with the following physics formula:

$$Voltage\ =\ Current\ \cdot\ Resistance$$

The two formulas are identical except that one applies to money and the other applies to electricity. So, imagine a strategy in which you adjust a circuit's current and resistance in line with a product's price and quantity. When you measure the resulting voltage, you can interpret this measurement as a value for the total revenue. You've reduced a problem in microeconomics to one that can be solved in an analog circuit.

This chapter describes a few alternative models for quantum computing. In each model, we use reducibility to change one kind of problem into another more quantum-ready problem.

Quantum simulation

Wouldn't it be nice if we could move current through materials with no resistance at all? The implications for the world's energy needs would be enormous. We could deliver energy to homes with no loss along the way. Batteries would be able to store their charges for indefinite amounts of time.

In fact, we have created electricity with no resistance, but the only materials that we know can do this require temperatures near absolute zero. The energy we need to cool these materials to very low temperatures far outweighs the benefits we'd gain in any widely used applications. So, for now, our techniques for dealing with electricity involve lots of wasteful energy loss.

The holy grail for electricity would be a **high-temperature superconductor** – a kind of material that conducts electricity with no resistance at temperatures that are practical for everyday use. We know a lot about the physics behind superconductivity. We have a formula called the **Hubbard Hamiltonian**, which governs the actions of electrons inside a superconductor. The problem is we haven't yet found a kind of material that behaves correctly at sufficiently high temperatures.

We can look for solutions to the Hubbard Hamiltonian formula, but that formula involves spin densities, harmonic oscillators, and other such terms. Starting with the Hubbard Hamiltonian formula, no one's found a way to "solve for x." We've tried to find approximate solutions using classical computers but with a material's electrons in superposition, the amount of classical processing resources we need grows exponentially. The world's largest computers can't handle the load!

However, there's hope on the horizon. A certain kind of quantum computing offers a promising strategy. We start by creating an array of standing laser waves. Atoms move from place to place within the array, and their movement is governed by the Hubbard Hamiltonian. We can observe this movement and interpret it as a model for the behavior of superconducting materials. Thus, the problem of analyzing superconducting materials is reducible to problems concerning standing laser beams.

Superposition and entanglement play roles in a laser array, so the technology being used in this scenario is a form of quantum computing. Unlike gate-based computing, we're not starting with circuits and gates, but we're still using hardware's quantum mechanical properties to solve certain kinds of problems. Quantum simulation is a promising field for business and research.

Quantum annealing

With rising concerns about climate change, it's useful to think about the way temperature varies by geographic location. *Figure 10.1* has a fictional map of temperatures in a particular region.

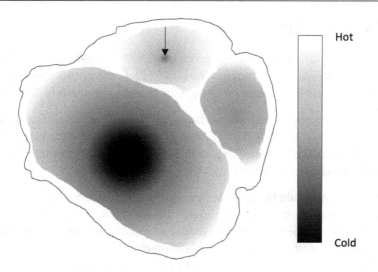

Figure 10.1 – A temperature map

Our goal is to find the coldest spot on this map. One strategy would be to start at an arbitrary point and work our way to points of ever-decreasing temperature. This is called a **greedy method** because the algorithm always moves in a direction that yields a better value.

The arrow in *Figure 10.1* shows a possible path using the greedy approach. The arrow starts at the top of the map, moves downward until it hits the center of the northernmost region, and then stops. The arrow stops because, having reached this point, movement in any direction would raise the temperature. The algorithm has found a **local extreme point** without ever getting near the coldest spot on the map.

Annealing is a way of moving slowly toward and less directly toward the goal in order to maximize the chance of finding the system's **global extreme point**, such as the coldest point on a temperature map. The word *annealing* started as a method in materials science for treating metals and was borrowed by computer scientists as the name for a classical computing algorithm. When programmers use **simulated annealing**, they start the search of a map by allowing some movements that go to higher-temperature locations. Roughly speaking, a movement is allowed if the percentage by which it raises the temperature is within a certain randomly selected limit. As the search progresses, the range of allowable movements becomes more and more narrow. As the search nears its end, movements go only toward colder places.

With hardware that does **quantum annealing**, a collection of entangled 5,000 qubits changes its state based on magnetic forces that coax the qubits in one direction or another. The application of these magnetic forces (and other effects) mimics the shape of whatever map is being studied. With the superposition of qubit states, the system represents exponentially many possibilities all at the same time.

When the process is finished, the system has settled into a low-energy state, and the qubits have become classical zero and one bits. At this point, we read the bits and interpret them as a representation of our desired result.

Figure 10.2 illustrates an important element of the success of quantum annealing. In this figure, the curve represents the temperatures as we move along some path on a temperature map. On the left side of the figure, an object is stuck inside a local extreme point. With the greedy method, the object never leaves this point and never finds the global extreme point. As a classical physical system, this is analogous to the object's being immobilized by the mountainous slopes on either side of it.

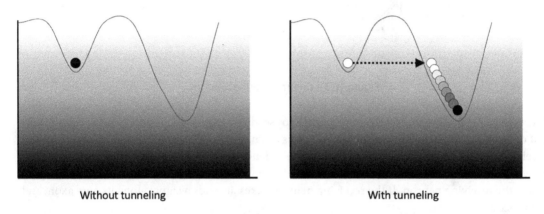

Without tunneling   With tunneling

Figure 10.2 – Quantum tunneling

However, the laws of quantum mechanics include a phenomenon called **tunneling**. With tunneling, an energy barrier that would be insurmountable in classical physics has a non-zero probability of being crossed. So, on the right side of *Figure 10.2*, the object moves out of its rut where it finds a better extreme point.

It's important to remember that maps and curves of this kind can apply to all kinds of problems – not only problems that involve two-dimensional space. You can reduce the value of a company's stock to a four-dimensional space with interest rate, employment rate, rate of inflation, and consumer confidence as the space's four dimensions. You can reduce a customer's preference for music to a three-dimensional map whose axes are genre, year, and recording quality. The possibilities are endless.

Quantum neural nets

Your brain is made up of approximately 200 billion cells, of which about half are **neurons**. *Figure 10.3* illustrates an interaction between two neurons.

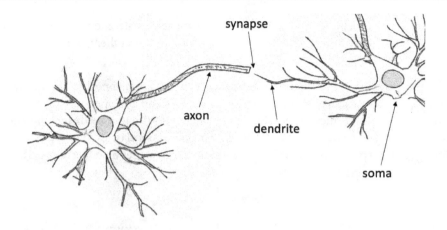

Figure 10.3 – Communicating neurons

A neuron communicates with a neighboring neuron by sending a chemical substance (a **neurotransmitter**) out of its **axon**. The neurotransmitter leaps across a **synapse** – a little gap between the sending and receiving neuron. On the other side of the synapse, a **dendrite** receives neurotransmitter and then forwards a signal to the receiving neuron's **soma**. Receipt of this signal may cause an electrical **spike** inside the receiving neuron. If the receiving neuron spikes, it sends a signal along its own axon, and the process continues.

In 1943, researchers named McCulloch and Pitts [1] described an electrical device whose behavior modeled that of a neuron. An **artificial neuron** has variable weights, as shown in *Figure 10.4*.

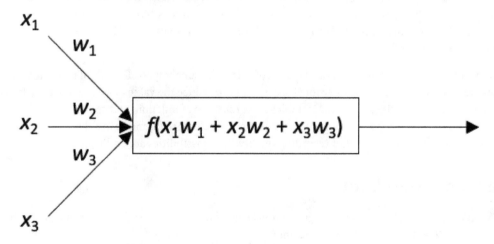

Figure 10.4 – An artificial neuron

In *Figure 10.4*, the input lines on the left correspond to a biological neuron's dendrites. The output line on the right represents a bio neuron's axon. The formula in the middle computes a value based on the inputs the unit receives, with each input x_i multiplied by a particular weight w_i. If the value of f is above a certain threshold, the artificial neuron spikes and sends a signal along its own axon.

In 1958, psychologist Frank Rosenblatt used the McCulloch-Pitts model as a building block for his **Perceptron** [2] – a device that could learn by adjusting the weights of its artificial neurons. You'd feed input into the Perceptron's outermost dendrites. Values would trickle through the web of artificial neurons, and output would appear from the axons at the other end. Based on some measure of the validity of the output, the Perceptron's weights would be tweaked, and more input would be supplied. After many rounds of input and output, the Perceptron's weights would gravitate toward a state in which the system yielded more and more valid outputs. In a way, the system learned.

The Perceptron and devices like it came to be known as **neural nets**. For many years, neural nets were second-class citizens in the world of artificial intelligence, but in 2012, Krizhevsky, Sutskever, and Hinton published a paper [3] that brought neural nets to the forefront. Since then, much of the work in machine learning has been about creating better and better neural nets.

So what does this have to do with quantum computing? Think about a quantum circuit containing R_X and R_Y gates. *Chapter 3* tells us that we can rotate the Bloch sphere by any angle. For example, consider these statements:

```
angle = math.pi / 2
circ.ry(angle, 0)
```

They rotate a qubit $\frac{\pi}{2}$ radians around the Bloch sphere's y axis. A classical neural net learns by adjusting the weights of its artificial neurons. Instead of adjusting a neuron's weights, we can adjust the angles of rotation in a quantum circuit. After many rounds of input and output, the quantum neural net's rotation angles gravitate toward values that represent correct answers.

In this decade, artificial intelligence is poised to change the world as we know it. There are several ways to combine quantum computing with machine learning. Quantum neural nets represent only one of many different approaches.

Solving unsolvable problems

In 1936, Alan Turing wrote a landmark paper [4] in which he showed that a particular well-defined mathematical problem is impossible to solve. In this case, the word "impossible" means that no classical algorithm running for a finite amount of time will ever reach a final, concluding step. To prove his claim, Turing created a precise definition of what it means to be an algorithm and went on to devise a problem that contains its own circular knot.

Consider the following sentence:

Is "no" the correct answer to the question that you're reading right now?

This sentence is like a double-edged sword:

- If you answer "yes" to the sentence, you're saying "Yes. It's true that "no" is the correct answer." But, if "no" is really the correct answer, your utterance of the word "yes" is incorrect.

- If you answer "no" to the sentence, you're saying "No. It's not true that "no" is the correct answer." But, if "no" isn't the correct answer, your reply with the word "no" is incorrect.

Either way, you're wrong. You just can't win.

In the same way, Turing constructed a mathematical problem whose solution, with a classical algorithm in a finite number of steps, would be self-contradictory.

Since then, dozens of such problems have been discovered. Here are a few of them:

- Given an arbitrary polynomial equation with integer coefficients and a finite number of variables, decide whether there are integer values for the variables that satisfy the equation. [5]

- Given a set of dominos, each having two strings of characters (instead of two halves containing dots), decide whether there's a way to line up the dominos, with repetitions permitted, so that the combined string on the top row is identical to the combined string on the bottom row. [6]

- Given the rules used by airline companies to set the pricing of flights, find the least expensive way to get from one airport to another. [7]

- Given a physical system, calculate the difference between its lowest energy state (its **ground state**) and its next-to-lowest energy state (its **first excited state**). [8]

Figure 10.5 shows a Helium atom with electrons in both the ground state and the first excited state. Electrons are talented things. They can jump from one state to another. When the electron in the first excited state jumps down to the ground state, it emits an amount of energy equal to the difference in energies between those two states. We can measure this energy by passing it through a prism and observing the spectrum that comes out of the prism. That's where this energy difference gets its name. The difference in energies between a system's ground state and its first excited state is called its **spectral gap**.

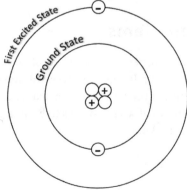

Figure 10.5 – A Helium atom

In their 2015 paper [8], Cubitt, Perez-Garcia, and Wolf proved that calculating a system's spectral gap is, in general, unsolvable. This raises an exciting possibility.

Someday, in the not-too-distant future, you want to travel from London to Bangalore. Being a skinflint like Barry Burd, you aim to find the very cheapest itinerary. You check the conventional travel sites, and they all come up with similar numbers – about 2,000 GBP.

However, then you stumble on the site `cheapprovider.com`, which has access to a new kind of quantum computer. Your airline pricing problem is reducible to some physical system's spectral gap calculation. Both problems are unsolvable by Alan Turing's reckoning, but the equations governing your pricing problem and this particular spectral gap calculation are surprisingly similar. A solution to the spectral gap problem can be reinterpreted as a solution to your airline question.

When you submit your travel request to `cheapprovider.com`, the site sends the spectral gap problem to a real physical system. The system's equipment measures the spectral gap and reports the result back to `cheapprovider.com`. The *cheapprovider* site reinterprets this result as a solution to your travel query and sends the itinerary to you. In the end, you save 400 GBP.

The story you've just read is science fiction. We can find approximate solutions to unsolvable problems, but exact solutions are never guaranteed. No computing device in existence today can do the things that I've described in these paragraphs, but the concepts are all valid. In principle, this scheme might work.

More speculatively, we can ask what secret sauce in physical reality allows it to do things that our smartest classical computers will never be able to do. Turing's original definition of the limits of computation is very convincing. The definition doesn't depend on the speed of a processor, the amount of memory, or any such limitation. I've often told my students that if aliens come to Earth and tell us that they've solved Alan Turing's famous unsolvable problem, they'll be lying. Of course, when I said this, I was thinking only about classical computing.

So, what does nature do that transcends step-by-step computation? The opinions voiced by experts vary a great deal from one another, but most authors' proposals include some mention of *infinity*. Whether it's infinite precision, an infinite amount of information, or infinitely many possible worlds, reality offers possibilities that we can only imagine. The nature of reality is ours to contemplate, ours to explore, and ours to discover.

Summary

Here you are in the last section of this book! I congratulate you on your effort (even if you skipped over many of the sections). Quantum computing isn't easy to understand, and you've taken the first step. You've explored algorithms using matrix algebra, and now you're ready for more.

There's no shortage of published information about quantum computing. With a quick search of the web, you can find blogs, videos, papers, and much more. For elegant descriptions of this book's algorithms, focus your attention on linear algebra. For insight into the science underlying quantum computing, read more about quantum physics. For information about the engineering challenges facing quantum computing professionals, learn about error detection and error correction. To practice writing quantum computing code, check the Qiskit documentation pages. And don't forget to investigate some other platforms for programming quantum computing algorithms. There's OpenQASM, Q#, Cirq, and Braket, to name just a few.

In writing this book, my primary goals were to inform and inspire. It's nice just to know things, but knowledge coupled with enthusiasm is even better. My interests are strongly entangled with quantum computing, and I hope your interests are headed in the same direction.

Finally, if you have questions about this book's material, visit `https://quantum.allmycode.com` where I post additional information as needed. You can also send an email to me at `quantum@allmycode.com`. I like hearing from readers. And yes . . . the reply that you receive will really be from me, Barry Burd.

References

[1] McCulloch, W.S., Pitts, W, (1943). *A logical calculus of the ideas immanent in nervous activity.* Bulletin of Mathematical Biophysics 5, pp. 115-133 `https://doi.org/10.1007/BF02478259`.

[2] Rosenblatt, F. (1958). *The perceptron: A probabilistic model for information storage and organization in the brain.* Psychological Review, 65(6), pp. 386-408. https://doi.org/10.1037/h0042519.

[3] Krizhevsky, A., Sutskever, I., Hinton, G.E. (2017-05-24). *ImageNet classification with deep convolutional neural networks* (PDF) available at `https://proceedings.neurips.cc/paper_files/paper/2012/file/c399862d3b9d6b76c8436e924a68c45b-Paper.pdf`. Communications of the ACM. 60 (6): pp. 84-90. doi:`10.1145/3065386`. ISSN `0001-0782`. S2CID `195908774`.

[4] Turing, A. M. (1937). *On Computable Numbers, with an Application to the Entscheidungsproblem*, Proceedings of the London Mathematical Society, Volume s2-42, Issue 1, pp. 230-265, `https://doi.org/10.1112/plms/s2-42.1.230`.

[5] Davis, M. (1973). *Hilbert's Tenth Problem is Unsolvable*, American Mathematical Monthly, vol.80, pp. 233-269.

[6] Post, E. L. (1946). *A variant of a recursively unsolvable problem* (PDF). Bull. Amer. Math. Soc. 52 (4): 264-269. doi:`10.1090/s0002-9904-1946-08555-9`. S2CID `122948861`.

[7] de Marcken, C. *Computational Complexity of Air Travel Planning.* `http://www.demarcken.org/carl/papers/ITA-software-travel-complexity/ITA-software-travel-complexity.pdf`

[8] Cubitt, T., Perez-Garcia, D., Wolf, M. (2015). *Undecidability of the spectral gap.* Nature 528, pp. 207-211. `https://doi.org/10.1038/nature16059`.

Assessments

Chapter 1, New Ways to Think about Bits

1.
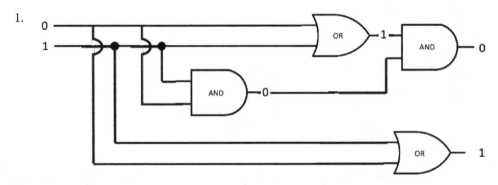

2. $1 \cdot 128 + 0 \cdot 64 + 1 \cdot 32 + 1 \cdot 16 + 1 \cdot 8 + 0 \cdot 4 + 0 \cdot 2 + 1 \cdot 1 = 185$

3. $1 \cdot 5 + 0 \cdot 8 + 3 \cdot 4 + 2 \cdot 4 = 25$

4.
$$\begin{pmatrix} 7 & 17 & 5 \\ 14 & -9 & 7 \end{pmatrix}$$

5.
$$\begin{pmatrix} 2 \begin{pmatrix} 8 & 4 & 0 \\ 1 & 3 & 5 \end{pmatrix} \\ 3 \begin{pmatrix} 8 & 4 & 0 \\ 1 & 3 & 5 \end{pmatrix} \\ 1 \begin{pmatrix} 8 & 4 & 0 \\ 1 & 3 & 5 \end{pmatrix} \end{pmatrix} = \begin{pmatrix} 16 & 8 & 0 \\ 2 & 6 & 10 \\ 24 & 12 & 0 \\ 3 & 9 & 15 \\ 8 & 4 & 0 \\ 1 & 3 & 5 \end{pmatrix}$$

6.
$$\begin{pmatrix} 0 & 1 \\ 1 & 0 \end{pmatrix} \begin{pmatrix} 1 & 0 & 0 & 0 \\ 0 & 1 & 1 & 1 \end{pmatrix}$$

7.

$$\begin{pmatrix} 1 & 0 & 0 & 0 \\ 0 & 1 & 1 & 1 \end{pmatrix} \otimes \begin{pmatrix} 0 & 1 \\ 1 & 0 \end{pmatrix} = \begin{pmatrix} 1\begin{pmatrix} 0 & 1 \\ 1 & 0 \end{pmatrix} & 0\begin{pmatrix} 0 & 1 \\ 1 & 0 \end{pmatrix} & 0\begin{pmatrix} 0 & 1 \\ 1 & 0 \end{pmatrix} & 0\begin{pmatrix} 0 & 1 \\ 1 & 0 \end{pmatrix} \\ 0\begin{pmatrix} 0 & 1 \\ 1 & 0 \end{pmatrix} & 1\begin{pmatrix} 0 & 1 \\ 1 & 0 \end{pmatrix} & 1\begin{pmatrix} 0 & 1 \\ 1 & 0 \end{pmatrix} & 1\begin{pmatrix} 0 & 1 \\ 1 & 0 \end{pmatrix} \end{pmatrix} = \begin{pmatrix} 0 & 1 & 0 & 0 & 0 & 0 & 0 & 0 \\ 1 & 0 & 0 & 0 & 0 & 0 & 0 & 0 \\ 0 & 0 & 0 & 1 & 0 & 1 & 0 & 1 \\ 0 & 0 & 1 & 0 & 1 & 0 & 1 & 0 \end{pmatrix}$$

8.

```
import numpy as np

A = np.matrix( [[1, 2, 3, 0],
                [2, 1, -1, 3]] )
B = np.matrix( [[1, 1, -2],
                [3, 2, -1],
                [0, 4, 3],
                [3, -3, 5]] )
print(np.dot(A, B))
```

```
import numpy as np

A = np.matrix( [[2],
                [3],
                [1]] )
B = np.matrix( [[8, 4, 0],
                [1, 3, 5]] )
print(np.kron(A, B))
```

Chapter 2, What Is a Qubit?

1. 1,000.

2. 0.

3. 500.

4. Approximately 250.

5. The number of zeros is usually between 490 and 510. Very rarely does the number become lower than 480 or higher than 520.

6.

```
from qiskit import QuantumCircuit
circuit = QuantumCircuit(3, 4)
circuit.h(0)
circuit.h(2)
circuit.barrier()
circuit.measure([1], [2])
circuit.measure([2], [1])
circuit.measure([0], [3])
display(circuit.draw('latex'))
```

Chapter 3, Math for Qubits and Quantum Gates

1. A. $\dfrac{1}{\sqrt{7}}\begin{pmatrix} \sqrt{3} \\ -2 \end{pmatrix}$ represents a qubit because $\left|\dfrac{\sqrt{3}}{\sqrt{7}}\right|^2 + \left|\dfrac{-2}{\sqrt{7}}\right|^2 = \dfrac{3}{7} + \dfrac{4}{7} = 1$.

    B. $\begin{pmatrix} 1 \\ 1 \end{pmatrix}$ doesn't represent a qubit because $\left|1\right|^2 + \left|1\right|^2 = 2 \neq 1$.

    C. $\begin{pmatrix} 1 \\ 0 \\ 0 \end{pmatrix}$ doesn't represent a qubit because it has three entries, not two.

2. Here is the result:

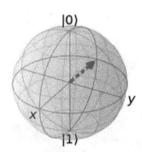

3. $(Z^T)\,Z = \begin{pmatrix} 1 & 0 \\ 0 & -1 \end{pmatrix}\begin{pmatrix} 1 & 0 \\ 0 & -1 \end{pmatrix} = \begin{pmatrix} 1 & 0 \\ 0 & 1 \end{pmatrix} = I$

4. $Z\,|+\rangle = \begin{pmatrix} 1 & 0 \\ 0 & -1 \end{pmatrix}\begin{pmatrix} \dfrac{1}{\sqrt{2}} \\ \dfrac{1}{\sqrt{2}} \end{pmatrix} = \begin{pmatrix} \dfrac{1}{\sqrt{2}} \\ -\dfrac{1}{\sqrt{2}} \end{pmatrix} = |-\rangle$

5. Here is the Qiskit code:

```
qc = QuantumCircuit(1)
qc.h(0)
qc.z(0)
```

6. $\left(R_Y\left(\dfrac{\pi}{2}\right)^T\right)R_Y\left(\dfrac{\pi}{2}\right) = \dfrac{1}{\sqrt{2}}\begin{pmatrix} 1 & 1 \\ -1 & 1 \end{pmatrix}\dfrac{1}{\sqrt{2}}\begin{pmatrix} 1 & -1 \\ 1 & 1 \end{pmatrix} = \dfrac{1}{2}\begin{pmatrix} 2 & 0 \\ 0 & 2 \end{pmatrix} = I$

7. According to *Figure 3.25*, $\sin\dfrac{\pi}{4} = \cos\dfrac{\pi}{4} = \dfrac{1}{\sqrt{2}}$. So, we have
.

$$R_Y\left(\dfrac{\pi}{4}\right) = \begin{pmatrix} \cos\dfrac{\pi}{4} & -\sin\dfrac{\pi}{4} \\ \sin\dfrac{\pi}{4} & \cos\dfrac{\pi}{4} \end{pmatrix} = \begin{pmatrix} \dfrac{1}{\sqrt{2}} & -\dfrac{1}{\sqrt{2}} \\ \dfrac{1}{\sqrt{2}} & \dfrac{1}{\sqrt{2}} \end{pmatrix} = \dfrac{1}{\sqrt{2}}\begin{pmatrix} 1 & -1 \\ 1 & 1 \end{pmatrix}$$

8. $H\,|0\rangle = \dfrac{1}{\sqrt{2}}\begin{pmatrix} 1 & 1 \\ 1 & -1 \end{pmatrix}\begin{pmatrix} 1 \\ 0 \end{pmatrix} = \dfrac{1}{\sqrt{2}}\begin{pmatrix} 1 \\ 1 \end{pmatrix}$

$R_Y\left(\dfrac{\pi}{2}\right)|0\rangle = \dfrac{1}{\sqrt{2}}\begin{pmatrix} 1 & -1 \\ 1 & 1 \end{pmatrix}\begin{pmatrix} 1 \\ 0 \end{pmatrix} = \dfrac{1}{\sqrt{2}}\begin{pmatrix} 1 \\ 1 \end{pmatrix}$

9. Here's the Hadamard gate acting on $|+\rangle$:

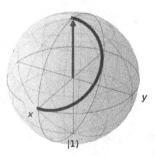

Here's the $R_Y\left(\dfrac{\pi}{2}\right)$ gate acting on $|+\rangle$:

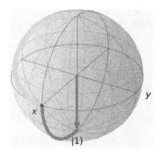

10. $XH = \begin{pmatrix} 0 & 1 \\ 1 & 0 \end{pmatrix} \dfrac{1}{\sqrt{2}} \begin{pmatrix} 1 & 1 \\ 1 & -1 \end{pmatrix} = \dfrac{1}{\sqrt{2}} \begin{pmatrix} 1 & -1 \\ 1 & 1 \end{pmatrix} = R_Y\left(\dfrac{\pi}{2}\right)$

Chapter 4, Qubit Conspiracy Theories

1. q_1 is 1, q_0 is 0.

2. You can demonstrate that writing $|\Phi^+\rangle$ in the form $\begin{pmatrix} a_0 \\ a_1 \end{pmatrix} \otimes \begin{pmatrix} b_0 \\ b_1 \end{pmatrix}$ is impossible. Here's how:

    Assume (to the contrary) that $\begin{pmatrix} 1 \\ 0 \\ 0 \\ 1 \end{pmatrix} = \begin{pmatrix} a_0 \\ a_1 \end{pmatrix} \otimes \begin{pmatrix} b_0 \\ b_1 \end{pmatrix}$. Then, $\begin{pmatrix} 1 \\ 0 \\ 0 \\ 1 \end{pmatrix} = \begin{pmatrix} a_0 b_0 \\ a_0 b_1 \\ a_2 b_0 \\ a_1 b_1 \end{pmatrix}$.

    But since $a_0 b_1 = 0$, either $a_0 = 0$ or $b_1 = 0$.

    In the first case, $a_0 b_0$ must be 0 instead of 1. In the second case, $a_1 b_1$ must be 0 instead of 1. Either way, things don't work out.

3.
$$(Z \otimes I)\,CNOT\,(I \otimes H)\,|00\rangle = \begin{pmatrix} 1 & 0 & 0 & 0 \\ 0 & 1 & 0 & 0 \\ 0 & 0 & -1 & 0 \\ 0 & 0 & 0 & -1 \end{pmatrix}\begin{pmatrix} 1 & 0 & 0 & 0 \\ 0 & 0 & 0 & 1 \\ 0 & 0 & 1 & 0 \\ 0 & 1 & 0 & 0 \end{pmatrix}\dfrac{1}{\sqrt{2}}\begin{pmatrix} 1 & 1 & 0 & 0 \\ 1 & -1 & 0 & 0 \\ 0 & 0 & 1 & 1 \\ 0 & 0 & 1 & -1 \end{pmatrix}\begin{pmatrix} 1 \\ 0 \\ 0 \\ 0 \end{pmatrix} = \dfrac{1}{\sqrt{2}}\begin{pmatrix} 1 \\ 0 \\ 0 \\ -1 \end{pmatrix} = |\Phi^-\rangle$$

4.
```
from qiskit import QuantumCircuit

circ = QuantumCircuit(2)
circ.h(0)
circ.cnot(0, 1)
circ.x(1)
circ.z(1)
display(circ.draw('latex', initial_state=True))
```

The resulting sphere looks like this:

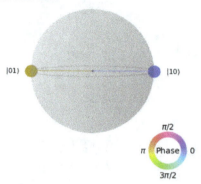

The $|01\rangle$ qubit's phase is π, and the $|10\rangle$ qubit's phase is 0.

5. $P(\,2\ girls\ +\ 1\ boy\,) = P(\,girl\ girl\ boy\,) + P(\,girl\ boy\ girl\,) + P(\,boy\ girl\ girl\,)$

$$= \frac{100}{205} \cdot \frac{100}{205} \cdot \frac{105}{205} + \frac{100}{205} \cdot \frac{105}{205} \cdot \frac{100}{205} + \frac{105}{205} \cdot \frac{100}{205} \cdot \frac{100}{205} = \frac{3 \cdot 100 \cdot 100 \cdot 105}{205^3}$$

6. There are eight ways to assign up-or-down arrows to the three measuring directions on the left. For the six ways that resemble the scenario in *Figure 4.35*, the probability of disagreement is $\frac{5}{9}$.

    For the remaining two ways, the probability of disagreement is 1 (see the following diagram):

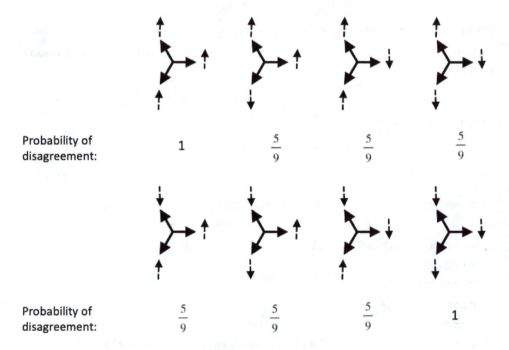

Probability of disagreement:	1	$\frac{5}{9}$	$\frac{5}{9}$	$\frac{5}{9}$

Probability of disagreement:	$\frac{5}{9}$	$\frac{5}{9}$	$\frac{5}{9}$	1

Each of these eight ways to assign up-or-down arrows occurs with a probability of $\frac{1}{8}$. So, the overall probability that the measurements disagree is this:

$$6\left(\frac{1}{8}\cdot\frac{5}{9}\right) + 2\left(\frac{1}{8}\cdot 1\right) = \frac{30}{72} + \frac{2}{8} = \frac{48}{72} = \frac{6}{9}$$

Chapter 5, A Fanciful Tale about Cryptography

1. Send information about any errors that you find to quantum@allmycode.com.

2. Send information about any errors that you find to quantum@allmycode.com.

3. If Eve measures Alice's qubits, then Alice and Bob will know about it. So, Alice and Bob won't use any of those qubits as their encryption key. So, if Eve knows some of Alice's zeros and ones, that information won't be useful to her.

4. The following example illustrates the use of the right-distributive law for tensor products:

$$\left(\begin{pmatrix}3 & 4\\5 & 6\end{pmatrix} + \begin{pmatrix}7 & 8\\9 & 0\end{pmatrix}\right) \otimes \begin{pmatrix}1\\2\end{pmatrix} = \begin{pmatrix}10 & 12\\14 & 6\end{pmatrix} \otimes \begin{pmatrix}1\\2\end{pmatrix} = \begin{pmatrix}10 & 12\\20 & 24\\14 & 6\\28 & 12\end{pmatrix}$$

$$\left(\begin{pmatrix}3 & 4\\5 & 6\end{pmatrix} \otimes \begin{pmatrix}1\\2\end{pmatrix}\right) + \left(\begin{pmatrix}7 & 8\\9 & 0\end{pmatrix} \otimes \begin{pmatrix}1\\2\end{pmatrix}\right) = \begin{pmatrix}3 & 4\\6 & 8\\5 & 6\\10 & 12\end{pmatrix} + \begin{pmatrix}7 & 8\\14 & 16\\9 & 0\\18 & 0\end{pmatrix} = \begin{pmatrix}10 & 12\\20 & 24\\14 & 6\\28 & 12\end{pmatrix}$$

5. When we cross-multiply, we use both the left- and right-distributive laws. In the following equations, one step uses the right-distributive law:

$$\underbrace{(\,\alpha\,|0\rangle + \beta\,|1\rangle\,)}_{(\ x\ +\ y\ )\otimes}\underbrace{(\,\alpha\,|0\rangle + \beta\,|1\rangle\,)}_{z}$$

$$= \underbrace{\alpha\,|0\rangle}_{x\ \otimes}\underbrace{(\,\alpha\,|0\rangle + \beta\,|1\rangle\,)}_{z} + \underbrace{\beta\,|1\rangle}_{+\ y\ \otimes}\underbrace{(\,\alpha\,|0\rangle + \beta\,|1\rangle\,)}_{z}$$

And the next step uses the left-distributive law:

$$\underbrace{\alpha\,|0\rangle}_{x}\,\otimes(\underbrace{\alpha\,|0\rangle + \beta\,|1\rangle}_{y}) + \underbrace{\beta\,|1\rangle}_{}(\underbrace{\alpha\,|0\rangle + \beta\,|1\rangle}_{z})$$

$$= \underbrace{\alpha\,|0\rangle\alpha\,|0\rangle}_{x\,\otimes\,y} + \underbrace{\alpha\,|0\rangle\beta\,|1\rangle}_{} + \underbrace{\beta\,|1\rangle\alpha\,|0\rangle}_{x\,\otimes\,z} + \underbrace{\beta\,|1\rangle\beta\,|1\rangle}_{}$$

We use the scalar multiplication law when we combine two qubits' scalars. Here's an example:

$$\alpha\,|0\rangle\,\beta\,|1\rangle = \alpha\begin{pmatrix}1\\0\end{pmatrix} \otimes \beta\begin{pmatrix}0\\1\end{pmatrix} = \begin{pmatrix}\alpha\\0\end{pmatrix} \otimes \begin{pmatrix}0\\\beta\end{pmatrix} = \begin{pmatrix}0\\\alpha\beta\\0\\0\end{pmatrix}$$

$$\alpha\beta(|0\rangle|1\rangle) = \alpha\beta\left(\begin{pmatrix}1\\0\end{pmatrix} \otimes \begin{pmatrix}0\\1\end{pmatrix}\right) = \alpha\beta\begin{pmatrix}0\\1\\0\\0\end{pmatrix} = \begin{pmatrix}0\\\alpha\beta\\0\\0\end{pmatrix}$$

$$(\alpha\beta|0\rangle)|1\rangle = \left(\alpha\beta\begin{pmatrix}1\\0\end{pmatrix}\right) \otimes \begin{pmatrix}0\\1\end{pmatrix} = \begin{pmatrix}\alpha\beta\\0\end{pmatrix} \otimes \begin{pmatrix}0\\1\end{pmatrix} = \begin{pmatrix}0\\\alpha\beta\\0\\0\end{pmatrix}$$

6. You can code the circuit as follows:

```
from qiskit import QuantumRegister, QuantumCircuit
from math import pi
import random
alice_q = QuantumRegister(1, 'alice_q')
bob_q = QuantumRegister(1, 'bob_q')
circ = QuantumCircuit(alice_q, bob_q)
circ.ry(pi/(random.uniform(2, 20)), alice_q[0])
circ.cnot(0, 1)
circ.measure_all()
display(circ.draw('latex'))
```

(In this code, the number 20 is my arbitrary choice.)

The circuit's matrix arithmetic is

$$CNOT\left(\begin{pmatrix} 1 \\ 0 \end{pmatrix} \otimes \begin{pmatrix} \alpha \\ \beta \end{pmatrix}\right) = \begin{pmatrix} 1 & 0 & 0 & 0 \\ 0 & 0 & 0 & 1 \\ 0 & 0 & 1 & 0 \\ 0 & 1 & 0 & 0 \end{pmatrix} \begin{pmatrix} \alpha \\ \beta \\ 0 \\ 0 \end{pmatrix} = \begin{pmatrix} \alpha \\ 0 \\ 0 \\ \beta \end{pmatrix} = \alpha|00\rangle + \beta|11\rangle$$

Both Alice's and Bob's qubits have $|\alpha|^2$ probability of being 0 and $|\beta|^2$ probability of being 1. So, in some sense, Alice's and Bob's qubits are the same, so you might think that this circuit clones Alice's qubit. But these qubits are entangled. True cloning would mean ending up with two qubits whose amplitudes are independent of one another. Instead of having $\alpha|00\rangle + \beta|11\rangle$ you'd have $(\alpha|0\rangle + \beta|1\rangle)(\alpha|0\rangle + \beta|1\rangle)$, which is $\alpha^2|00\rangle + \alpha\beta|01\rangle + \alpha\beta|10\rangle + \beta^2|11\rangle$.

7. In the following code, Eve randomly decides whether or not she'll apply Hadamard gates:

```
def setup_eve(circ):
    bob_q = circ.qubits[1]
    eve_c = circ.clbits[1]

    has_had = random.getrandbits(1)
    circ.barrier()
    if has_had:
        circ.h(bob_q)
    circ.measure(bob_q, eve_c)
    if has_had:
        circ.h(bob_q)
    circ.barrier()
    return circ

def had_agreement(circ):
    gate_counts = circ.count_ops()
    return not ('h' in gate_counts and gate_counts['h'] %
2 == 1)
```

8. Eve can entangle qubits so that they're both in the same state, and Eve will get the same value as Bob when she measures her qubit. But that doesn't guarantee that this measurement will agree with the bit that Alice intended to send.

    For example, imagine that Alice sends a $|0\rangle$ without applying a Hadamard gate. After Eve applies entanglement, Eve has two qubits that either both measure as zeros or both measure as ones.

Eve sends one of them onward to Bob. If both qubits measure as ones, then Eve doesn't have an accurate record of Alice's randomly generated bits. In addition, Bob's measurement of 1 doesn't agree with Alice's 0, so Alice and Bob detect Eve's unwanted presence.

Chapter 6, Quantum Networking and Teleportation

1. In the *Quantum operations for teleportation* section, we derive the following formula:

$$\frac{1}{2}[(\alpha|0\rangle + \beta|1\rangle)|00\rangle + (\alpha|0\rangle - \beta|1\rangle)|01\rangle + (\beta|0\rangle + \alpha|1\rangle)|10\rangle + (-\beta|0\rangle + \alpha|1\rangle)|11\rangle]$$

When the initial state of Alice's qubit is $|0\rangle$, $\alpha = 1$ and $\beta = 0$. So, the formula becomes

$$\frac{1}{2}[\,|0\rangle|00\rangle + |0\rangle|01\rangle + |1\rangle|10\rangle + |1\rangle|11\rangle\,]$$

We have four possibilities, and each possibility has three qubits. Alice measures the middle qubit. When that measurement yields the value 1, Bob applies the X gate. So, we have

$$\frac{1}{2}[\,|0\rangle|00\rangle + |0\rangle|01\rangle + (X|1\rangle)|10\rangle + (X|1\rangle)|11\rangle\,]$$

$$= \frac{1}{2}[\,|0\rangle|00\rangle + |0\rangle|01\rangle + (|0\rangle)|10\rangle + (|0\rangle)|11\rangle\,]$$

$$= |0\rangle\frac{1}{2}[\,|00\rangle + |01\rangle + |10\rangle + |11\rangle\,]$$

No matter what values Alice's measurements yield, Bob's qubit is in the $|0\rangle$ state. So the value of Alice's qubit has been teleported to Bob.

The calculation is similar when the initial state of Alice's qubit is $|1\rangle$.

2. When you hardcode alpha = 0.8228 and beta = 0.5683, a call to add_gates gives you an error message. The message is Sum of amplitudes-squared does not equal one. This happens because the value of $|0.8228|^2 + |0.5683|^2$ isn't close enough to 1. Instead, it's 0.99996473. Qiskit randomly generates alpha and beta values and displays those values in the circuit diagram with only four digits of accuracy, but if you execute print(alpha) and print(beta), you get more precise values like 0.8228104952966586 and 0.5683158354556622.

3. To code the circuit, change to functions as follows:

```
def create_registers():
    alice_q = QuantumRegister(1, 'alice (q)')
    bob_q = QuantumRegister(1, 'bob (q)')
    circ = QuantumCircuit(alice_q, bob_q)
    return circ
```

```
def add_gates(circ, alpha, beta):
    circ.initialize([alpha, beta], 0)
    circ.cnot(0, 1)
    circ.measure_all()
    return circ
```

The circuit's matrix arithmetic is

$$CNOT\left(\begin{pmatrix}1\\0\end{pmatrix}\otimes\begin{pmatrix}\alpha\\\beta\end{pmatrix}\right)=\begin{pmatrix}1&0&0&0\\0&0&0&1\\0&0&1&0\\0&1&0&0\end{pmatrix}\begin{pmatrix}\alpha\\\beta\\0\\0\end{pmatrix}=\begin{pmatrix}\alpha\\0\\0\\\beta\end{pmatrix}=\alpha|00\rangle+\beta|11\rangle$$

Both Alice's and Bob's qubits have a $|\alpha|^2$ probability of being 0 and a $|\beta|^2$ probability of being 1. So, in some sense, Alice's and Bob's qubits are the same, and you might think that this circuit teleports Alice's qubit. But these qubits are entangled. True cloning would mean that Alice's qubit ends up in state $|0\rangle$. Instead of having, $\alpha|00\rangle+\beta|11\rangle$, you'd have $\alpha|00\rangle+\beta|10\rangle$.

4. The circuit's matrix arithmetic is as follows:

$$(I\otimes I\otimes H)(I\otimes CNOT)\frac{1}{\sqrt{2}}(|00\rangle+|11\rangle)(\alpha|0\rangle+\beta|1\rangle)=$$

$$\left(\begin{pmatrix}1&0\\0&1\end{pmatrix}\otimes\begin{pmatrix}1&0\\0&1\end{pmatrix}\otimes\frac{1}{\sqrt{2}}\begin{pmatrix}1&1\\1&-1\end{pmatrix}\right)\left(\begin{pmatrix}1&0\\0&1\end{pmatrix}\otimes\begin{pmatrix}1&0&0&0\\0&0&0&1\\0&0&1&0\\0&1&0&0\end{pmatrix}\right)\left(\frac{1}{\sqrt{2}}\begin{pmatrix}1\\0\\0\\1\end{pmatrix}\otimes\begin{pmatrix}\alpha\\\beta\end{pmatrix}\right)=$$

$$\frac{1}{2}\begin{pmatrix}1&1&0&0&0&0&0&0\\1&-1&0&0&0&0&0&0\\0&0&1&1&0&0&0&0\\0&0&1&-1&0&0&0&0\\0&0&0&0&1&1&0&0\\0&0&0&0&1&-1&0&0\\0&0&0&0&0&0&1&1\\0&0&0&0&0&0&1&-1\end{pmatrix}\begin{pmatrix}1&0&0&0&0&0&0&0\\0&0&0&1&0&0&0&0\\0&0&1&0&0&0&0&0\\0&1&0&0&0&0&0&0\\0&0&0&0&1&0&0&0\\0&0&0&0&0&0&0&1\\0&0&0&0&0&0&1&0\\0&0&0&0&0&1&0&0\end{pmatrix}\begin{pmatrix}\alpha\\\beta\\0\\0\\0\\0\\\alpha\\\beta\end{pmatrix}=$$

$$\frac{1}{2}\begin{pmatrix}1&0&0&1&0&0&0&0\\1&0&0&-1&0&0&0&0\\0&1&1&0&0&0&0&0\\0&-1&1&0&0&0&0&0\\0&0&0&0&1&0&0&1\\0&0&0&0&1&0&0&-1\\0&0&0&0&0&1&1&0\\0&0&0&0&0&-1&1&0\end{pmatrix}\begin{pmatrix}\alpha\\\beta\\0\\0\\0\\0\\\alpha\\\beta\end{pmatrix}=\frac{1}{2}\begin{pmatrix}\alpha\\\alpha\\\beta\\-\beta\\\beta\\-\beta\\\alpha\\\alpha\end{pmatrix}=$$

$$\frac{1}{2}(\alpha|000\rangle+\alpha|001\rangle+\beta|010\rangle-\beta|011\rangle+\beta|100\rangle-\beta|101\rangle+\alpha|110\rangle+\alpha|111\rangle)$$

5.

```python
def create_registers():
    alice_q = QuantumRegister(1, 'alice (q)')
    peter_alice_q = QuantumRegister(1, 'peter/alice (q)')
    peter_bob_q = QuantumRegister(1, 'peter/bob (q)')
    bob_c = ClassicalRegister(3, 'bob (c)')
    pedro_bob_q = QuantumRegister(1, 'pedro/bob (q)')
    pedro_carol_q = QuantumRegister(1, 'pedro/carol (q)')
    carol_c = ClassicalRegister(3, 'carol (c)')
    circ = QuantumCircuit(alice_q, peter_alice_q,
                          peter_bob_q, bob_c,
                          pedro_bob_q, pedro_carol_q,
                          carol_c)
    return circ

def add_gates(circ, alpha, beta):
    circ.initialize([alpha, beta], 0)
    circ.barrier()
    circ.h(1)
    circ.cnot(1, 2)
    circ.barrier()
    circ.cnot(0, 1)
    circ.h(0)
    circ.barrier()
    circ.measure(0, 0)
    circ.measure(1, 1)
    with circ.if_test((1, 1)):
        circ.x(2)
    with circ.if_test((0, 1)):
        circ.z(2)
    circ.barrier()

    circ.h(3)
    circ.cnot(3, 4)
    circ.barrier()
    circ.cnot(2, 3)
```

```
    circ.h(2)
    circ.barrier()
    circ.measure(2, 3)
    circ.measure(3, 4)
    with circ.if_test((4, 1)):
        circ.x(4)
    with circ.if_test((3, 1)):
        circ.z(4)

    circ.measure(4, 5)
    return circ

result = job.result()
counts = result.get_counts(circ)
counts_m = marginal_counts(counts, [5])
number_of_0s = counts_m.get('0')
number_of_1s = counts_m.get('1')
alpha = np.sqrt(number_of_0s / shots)
beta = np.sqrt(number_of_1s / shots)
print("|\u03C8\u27E9 ({:.4f}, {:.4f})".format(alpha,
beta))
```

Chapter 7, Deutsch's Algorithm

1. This circuit is not reversible. When $x = y$, the output is 01 regardless of whether x and y are both 0 or x and y are both 1. When $x \neq y$, the output is 10 regardless of which input is 0 and which is 1.

2. The number of correct shots varies quite a lot because the amount of noise isn't predictable. Also, some quantum computers tend to be less noisy than others. On IBM devices, the number of correct shots out of 100 is typically in the 90s but sometimes in the 80s.

3. The CNOT gate affects both the top and bottom qubits. But when we add an X gate, we add it to the bottom qubit. The top qubit (the qubit that we eventually measure) remains unchanged.

4. This circuit implements the *Opposite_of* function. The leftmost X gate reverses the roles of inputs 0 and 1 from what they'd be with the *Identity* function. Then, the rightmost X gate restores the top qubit to its original 0 value or 1 value. That rightmost X gate has no effect on the topmost qubit's phase.

5. Here's the code for the new `get_oracle` function:

```python
def get_oracle(function):
    oracle = QuantumCircuit(2)
    # if function == SimpleBinary.ZERO:
        # Do nothing
    if function == SimpleBinary.ONE:
        oracle.x(1)
    elif function == SimpleBinary.SAME_AS:
        oracle.cnot(0, 1)
    elif function == SimpleBinary.OPPOSITE_OF:
        oracle.cnot(0, 1)
        oracle.x(1)
    return oracle
```

Here's the code that calls `get_oracle`:

```python
circ = QuantumCircuit(2, 1)
function = get_function()

circ.x(1)
circ.h(0)
circ.h(1)
circ.barrier()
oracle = get_oracle(function)
circ = circ.compose(oracle)
circ.barrier()
circ.h(0)
circ.measure(0, 0)
display(circ.draw('latex'))
```

6. There are 16 different ways to define a binary function $f(x_1, x_2)$. Here's why:

To define $f(x_1, x_2)$, you have to fill in four blanks:

$$f(0, 0) = \_$$
$$f(0, 1) = \_$$
$$f(1, 0) = \_$$
$$f(1, 1) = \_$$

There are two ways (0 or 1) for the first blank, two possibilities for the second, two possibilities for the third, and two possibilities for the fourth. All in all, that makes $2 \cdot 2 \cdot 2 \cdot 2 = 16$ possibilities.

There are only two ways to define constant functions. You can put 0 in all four blanks or put 1 in all four blanks.

There are six ways to define balanced functions. Here's why.

With a balanced function, two blanks are 0 and two blanks are 1. Once you've decided which two blanks get 0, you know for sure that the other two blanks will get 1. So, how many ways are there to choose the two blanks that get 0? Here are all six ways:

$f(0, 0) = 0$	0	0	1	1	1
$f(0, 1) = 0$	1	1	0	0	1
$f(1, 0) = 1$	0	1	0	1	0
$f(1, 1) = 1$	1	0	1	0	0

Of the 16 two-input binary functions, eight are either constant or balanced. The remaining eight are neither constant nor balanced.

7.   An oracle for $f(x_1, x_2) = 1$ is shown here:

An oracle for $f(x_1, x_2) = x_1$ is shown here:

An oracle for $f(x_1, x_2) = \neg x_2$ is shown here:

Chapter 8, Grover's Algorithm

1. It takes 20 qubits to search among 1,000,000 values because $2^{20} = 1,000,000$. The number of Grover iterations will be $\frac{\pi}{4}\sqrt{1000000} = \frac{\pi}{4}1000 \approx 0.7854 \cdot 1000 = 804 \cdot$

2.

$$\frac{1}{4}\begin{pmatrix} -3 & 1 & 1 & 1 & 1 & 1 & 1 & \\ 1 & -3 & 1 & 1 & 1 & 1 & 1 & 1 \\ 1 & 1 & -3 & 1 & 1 & 1 & 1 & 1 \\ 1 & 1 & 1 & -3 & 1 & 1 & 1 & 1 \\ 1 & 1 & 1 & 1 & -3 & 1 & 1 & 1 \\ 1 & 1 & 1 & 1 & 1 & -3 & 1 & 1 \\ 1 & 1 & 1 & 1 & 1 & 1 & -3 & 1 \\ 1 & 1 & 1 & 1 & 1 & 1 & 1 & -3 \end{pmatrix}$$

3. Here's the code:

```
import random

oracle_matrix = [
    [1, 0,  0, 0, 0, 0, 0, 0],
    [0, 1,  0, 0, 0, 0, 0, 0],
    [0, 0,  1, 0, 0, 0, 0, 0],
    [0, 0,  0, 1, 0, 0, 0, 0],
    [0, 0,  0, 0, 1, 0, 0, 0],
    [0, 0,  0, 0, 0, 1, 0, 0],
    [0, 0,  0, 0, 0, 0, 1, 0],
    [0, 0,  0, 0, 0, 0, 0, 1]
]

entry = random.randint(0, 7)
print(entry)
oracle_matrix[entry][entry] = -1

oracle = QuantumCircuit(3)
oracle.unitary(oracle_matrix, qubits=[0, 1, 2],
label='oracle')
oracle.barrier()
display(oracle.draw('latex'))
```

4. The expression is true only when x, y, and z are all false.

5. This run displays a matrix. The 4x4 upper left quadrant of the matrix represents the circuit's effect on the top two qubits. That part of the matrix is as follows:

$$\begin{pmatrix} \frac{1}{\sqrt{2}} & 0 & 0 & 0 \\ 0 & \frac{1}{\sqrt{2}} & 0 & 0 \\ 0 & 0 & \frac{1}{\sqrt{2}} & 0 \\ 0 & 0 & 0 & -\frac{1}{\sqrt{2}} \end{pmatrix}$$

When you apply this quadrant to a 4-element state vector, it puts a minus sign in front of the vector's $|11\rangle$ amplitude.

6. To answer this question, start with the code in the *Coding Grover's algorithm with high-level functions* section. Replace that code's `expression` assignment with the following line of code:

```
expression = '((m | t) & ~n) & ((t | n) & ~m)'
```

7. Let m stand for *The Identity Matrix*, t stand for *The Sixth Tensor*, and n stand for *The Jupyter Notebook*. If we allow the use of or, we get the following:

```
((m | t) &  ~n) &   ((t | n)  & ~m)
```

Before applying De Morgan's law, we must add double negations:

```
(~~(m | t)  & ~n) & (~~(t | n)  & ~m)
```

Then, we can apply De Morgan's law:

```
(~(~m & ~t) & ~n) & (~(~t & ~n) & ~m)
```

Here's the oracle for that Boolean expression:

```
from qiskit import QuantumRegister, QuantumCircuit
from qiskit.circuit.library.standard_gates import XGate

m = QuantumRegister(1, 'm')
t = QuantumRegister(1, 't')
n = QuantumRegister(1, 'n')
mt = QuantumRegister(1, 'not m and not t')
mtn = QuantumRegister(1, 'alice')
tn = QuantumRegister(1, 'not t and not n')
```

```
tnm = QuantumRegister(1, 'bob')
exp = QuantumRegister(1, 'alice and bob')
circ = QuantumCircuit(m, t, n, mt, mtn, tn, tnm, exp)


circ.h([0, 1, 2])


ctrl = XGate().control(2, ctrl_state='00')
circ.append(ctrl, qargs=[0, 1, 3])
circ.append(ctrl, qargs=[2, 3, 4])
circ.append(ctrl, qargs=[1, 2, 5])
circ.append(ctrl, qargs=[0, 5, 6])


circ.append(ctrl, qargs=[4, 6, 7])


circ.append(ctrl, qargs=[0, 5, 6])
circ.append(ctrl, qargs=[1, 2, 5])
circ.append(ctrl, qargs=[2, 3, 4])
circ.append(ctrl, qargs=[0, 1, 3])


circ.h([0, 1, 2])


display(circ.draw('latex'))
```

The oracle looks like this:

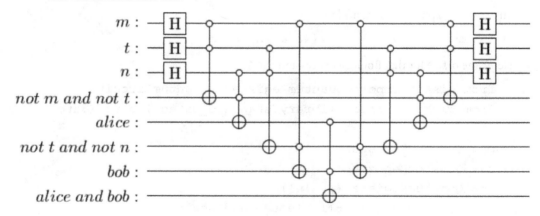

Chapter 9, Shor's Algorithm

1. We start by finding values of 3^n % 14:

$$3^0 \text{ \% } 14 = 1$$
$$3^1 \text{ \% } 14 = 3$$
$$3^2 \text{ \% } 14 = 9$$
$$3^4 \text{ \% } 14 = 11$$
$$3^5 \text{ \% } 14 = 5$$
$$3^6 \text{ \% } 14 = 1$$

Next, we calculate the following:

$$\left(3^{6/2} + 1\right)\left(3^{6/2} - 1\right) = 28 \cdot 26 = (2 \cdot 2 \cdot 7) \cdot (2 \cdot 13)$$

When we calculate 14/13, we find that it isn't an integer. But 14/2 = 7. So, 14 = 2·7.

2. We start by finding values of 2^n % 35:

$$2^0 \text{ \% } 35 = 1$$
$$2^1 \text{ \% } 35 = 2$$
$$2^2 \text{ \% } 35 = 4$$
$$2^3 \text{ \% } 35 = 8$$
$$2^4 \text{ \% } 35 = 16$$
$$2^5 \text{ \% } 35 = 32$$
$$2^6 \text{ \% } 35 = 29$$
$$2^7 \text{ \% } 35 = 23$$
$$2^8 \text{ \% } 35 = 11$$
$$2^9 \text{ \% } 35 = 22$$
$$2^{10} \text{ \% } 35 = 9$$
$$2^{11} \text{ \% } 35 = 18$$
$$2^{12} \text{ \% } 35 = 1$$

Next, we calculate the following:

$$\left(2^{12/2} + 1\right)\left(2^{12/2} - 1\right) = 65 \cdot 63 = (5 \cdot 13) \cdot (3 \cdot 3 \cdot 7)$$

When we calculate 35/13 and 35/3, we find that these aren't integers. But 35/5 = 7. So, 35 = 5·7.

3. Dividing 71 by 8 gives us 8 with a remainder of 7. So, $e^{\left(71\cdot\frac{\pi}{4}\right)}$ is the same as $e^{\left(7\cdot\frac{\pi}{4}\right)}$. According to *Figures 9.9* and *9.13*, $e^{\left(7\cdot\frac{\pi}{4}\right)}$ is equal to $\frac{1}{\sqrt{2}}-\frac{1}{\sqrt{2}}i$.

4. The QFT and QFT† matrices are shown here:

$$QFT = \begin{pmatrix} 1 & 1 & 1 & 1 \\ 1 & i & -1 & -i \\ 1 & -1 & 1 & -1 \\ 1 & -i & -1 & i \end{pmatrix} \qquad QFT^{\dagger} = \begin{pmatrix} 1 & 1 & 1 & 1 \\ 1 & -i & -1 & i \\ 1 & -1 & 1 & -1 \\ 1 & i & -1 & -i \end{pmatrix}$$

5. The 2 × 2 QFT matrix is the Hadamard matrix because the second roots of unity are 1 and -1.

6. The missing values are shown here:

$$\begin{pmatrix} \circ & \circ & \circ & \circ & \circ & \circ & \circ & \circ \\ \circ & \circ & \circ & \circ & \circ & \circ & \circ & \circ \\ \circ & \circ & \circ & \circ & \circ & \circ & \circ & \circ \\ \circ & \circ & \circ & \circ & \circ & \circ & \circ & \circ \\ \circ & \circ & \circ & \circ & \circ & \circ & \circ & \circ \\ \circ & \circ & \circ & \circ & \circ & \circ & \circ & \circ \\ \circ & \circ & \circ & \circ & \circ & \circ & \circ & \circ \\ \circ & \circ & \circ & \circ & \circ & \circ & \circ & \circ \end{pmatrix}\begin{pmatrix} 1 \\ 5 \\ 1 \\ 5 \\ 1 \\ 5 \\ 1 \\ 5 \end{pmatrix} = \begin{pmatrix} 1\cdot\circ + 5\cdot\circ + 1\cdot\circ + 5\cdot\circ + 1\cdot\circ + 5\cdot\circ + 1\cdot\circ + 5\cdot\circ \\ 1\cdot\circ + 5\cdot\circ + 1\cdot\circ + 5\cdot\circ + 1\cdot\circ + 5\cdot\circ + 1\cdot\circ + 5\cdot\circ \\ 1\cdot\circ + 5\cdot\circ + 1\cdot\circ + 5\cdot\circ + 1\cdot\circ + 5\cdot\circ + 1\cdot\circ + 5\cdot\circ \\ 1\cdot\circ + 5\cdot\circ + 1\cdot\circ + 5\cdot\circ + 1\cdot\circ + 5\cdot\circ + 1\cdot\circ + 5\cdot\circ \\ 1\cdot\circ + 5\cdot\circ + 1\cdot\circ + 5\cdot\circ + 1\cdot\circ + 5\cdot\circ + 1\cdot\circ + 5\cdot\circ \\ 1\cdot\circ + 5\cdot\circ + 1\cdot\circ + 5\cdot\circ + 1\cdot\circ + 5\cdot\circ + 1\cdot\circ + 5\cdot\circ \\ 1\cdot\circ + 5\cdot\circ + 1\cdot\circ + 5\cdot\circ + 1\cdot\circ + 5\cdot\circ + 1\cdot\circ + 5\cdot\circ \\ 1\cdot\circ + 5\cdot\circ + 1\cdot\circ + 5\cdot\circ + 1\cdot\circ + 5\cdot\circ + 1\cdot\circ + 5\cdot\circ \end{pmatrix}\begin{matrix} 0 \\ 1 \\ 2 \\ 3 \\ 4 \\ 5 \\ 6 \\ 7 \end{matrix}$$

$$= \begin{pmatrix} 1\cdot\circ + 5\cdot\circ + 1\cdot\circ + 5\cdot\circ + 1\cdot\circ + 5\cdot\circ + 1\cdot\circ + 5\cdot\circ \\ 1\!\!\!/\circ + 5\!\!\!/\circ + 1\!\!\!/\circ + 5\!\!\!/\circ + 1\!\!\!/\circ + 5\!\!\!/\circ + 1\!\!\!/\circ + 5\!\!\!/\circ \\ 1\!\!\!/\circ + 5\!\!\!/\circ + 1\!\!\!/\circ + 5\!\!\!/\circ + 1\!\!\!/\circ + 5\!\!\!/\circ + 1\!\!\!/\circ + 5\!\!\!/\circ \\ 1\!\!\!/\circ + 5\!\!\!/\circ + 1\!\!\!/\circ + 5\!\!\!/\circ + 1\!\!\!/\circ + 5\!\!\!/\circ + 1\!\!\!/\circ + 5\!\!\!/\circ \\ 1\cdot\circ + 5\cdot\circ + 1\cdot\circ + 5\cdot\circ + 1\cdot\circ + 5\cdot\circ + 1\cdot\circ + 5\cdot\circ \\ 1\!\!\!/\circ + 5\!\!\!/\circ + 1\!\!\!/\circ + 5\!\!\!/\circ + 1\!\!\!/\circ + 5\!\!\!/\circ + 1\!\!\!/\circ + 5\!\!\!/\circ \\ 1\!\!\!/\circ + 5\!\!\!/\circ + 1\!\!\!/\circ + 5\!\!\!/\circ + 1\!\!\!/\circ + 5\!\!\!/\circ + 1\!\!\!/\circ + 5\!\!\!/\circ \\ 1\!\!\!/\circ + 5\!\!\!/\circ + 1\!\!\!/\circ + 5\!\!\!/\circ + 1\!\!\!/\circ + 5\!\!\!/\circ + 1\!\!\!/\circ + 5\!\!\!/\circ \end{pmatrix}\begin{matrix} 0 \\ 1 \\ 2 \\ 3 \\ 4 \\ 5 \\ 6 \\ 7 \end{matrix}$$

$$= \begin{pmatrix} (\text{Some non-zero number.}) \\ 0 \\ 0 \\ 0 \\ (\text{Some non-zero number.}) \\ 0 \\ 0 \\ 0 \end{pmatrix}\begin{matrix} 0 \\ 1 \\ 2 \\ 3 \\ 4 \\ 5 \\ 6 \\ 7 \end{matrix}$$

7. In the coprime powers sequence for 22 (with coprime 3), the period is 5. But 5 isn't divisible by 2. You can't find $3^{5/2} + 1$ or $3^{5/2} - 1$:

$$3^0 \% 22 = 1$$
$$3^1 \% 22 = 3$$
$$3^2 \% 22 = 9$$
$$3^3 \% 22 = 5$$
$$3^4 \% 22 = 15$$
$$3^5 \% 22 = 1$$

8. Use repeated squaring to find $7^{11} \% 15$, like so:

$$7^{11} \% 15 = 7^{8 + 2 + 1} \% 15 = \left(7^8 \% 15\right)\left(7^2 \% 15\right)\left(7^1 \% 15\right) \% 15$$
$$= 1 \cdot 4 \cdot 7 \% 15 = 28 \% 15 = 13$$

9. The public key is $11 \cdot 13 = 143$.

The totient is $(11 - 1)(13 - 1) = 120$.

We find that $(7 \cdot 103) \% 120 = 1$, so the inverse coprime is 103.

So, to create the ciphertext, we find $97^7 \% 143$, which is 59.

To decrypt the ciphertext, we find $59^{103} \% 143$, which is 97.

10. As an example, set `phase` equal to `2*pi/3`. Then, the output from the code's `print` call might be `{ '0': 252, '1': 748 }`. With the number of 0 measurements close to 250, we apply the formula and get the following:

$$\text{phase} \approx \arccos\left(2\left(\frac{250}{1000}\right) - 1\right) = \arccos\left(-\frac{1}{2}\right) = \frac{2\pi}{3}$$

$$\text{phase} \approx \arccos\left(2\left(\frac{250}{1000}\right) - 1\right) = \arccos\left(-\frac{1}{2}\right) = \frac{2\pi}{3}$$

The formula gives us the phase of the bottom qubit.

Index

Packtpub.com

Subscribe to our online digital library for full access to over 7,000 books and videos, as well as industry leading tools to help you plan your personal development and advance your career. For more information, please visit our website.

Why subscribe?

- Spend less time learning and more time coding with practical eBooks and Videos from over 4,000 industry professionals

- Improve your learning with Skill Plans built especially for you

- Get a free eBook or video every month

- Fully searchable for easy access to vital information

- Copy and paste, print, and bookmark content

Did you know that Packt offers eBook versions of every book published, with PDF and ePub files available? You can upgrade to the eBook version at packtpub.com and as a print book customer, you are entitled to a discount on the eBook copy. Get in touch with us at customercare@packtpub.com for more details.

At www.packtpub.com, you can also read a collection of free technical articles, sign up for a range of free newsletters, and receive exclusive discounts and offers on Packt books and eBooks.

Other Books You May Enjoy

If you enjoyed this book, you may be interested in these other books by Packt:

A Practical Guide to Quantum Machine Learning and Quantum Optimization

Elías F. Combarro, Samuel González-Castillo

ISBN: 9781804613832

- Review the basics of quantum computing

- Gain a solid understanding of modern quantum algorithms

- Understand how to formulate optimization problems with QUBO

- Solve optimization problems with quantum annealing, QAOA, GAS, and VQE

- Find out how to create quantum machine learning models

- Explore how quantum support vector machines and quantum neural networks work using Qiskit and PennyLane

- Discover how to implement hybrid architectures using Qiskit and PennyLane and its PyTorch interface

Quantum Chemistry and Computing for the Curious

Keeper L. Sharkey, Alain Chancé

ISBN: 9781803243900

- Understand mathematical properties of the building blocks of matter
- Run through the principles of quantum mechanics with illustrations
- Design quantum gate circuit computations
- Program in open-source chemistry software packages such as Qiskit®
- Execute state-of-the-art-chemistry calculations and simulations
- Run companion Jupyter notebooks on the cloud with just a web browser
- Explain standard approximations in chemical simulations

Packt is searching for authors like you

If you're interested in becoming an author for Packt, please visit authors.packtpub.com and apply today. We have worked with thousands of developers and tech professionals, just like you, to help them share their insight with the global tech community. You can make a general application, apply for a specific hot topic that we are recruiting an author for, or submit your own idea.

Share Your Thoughts

Now you've finished *Quantum Computing Algorithms*, we'd love to hear your thoughts! Scan the QR code below to go straight to the Amazon review page for this book and share your feedback or leave a review on the site that you purchased it from.

https://packt.link/r/1-804-61737-7

Your review is important to us and the tech community and will help us make sure we're delivering excellent quality content.

Download a free PDF copy of this book

Thanks for purchasing this book!

Do you like to read on the go but are unable to carry your print books everywhere? Is your eBook purchase not compatible with the device of your choice?

Don't worry, now with every Packt book you get a DRM-free PDF version of that book at no cost.

Read anywhere, any place, on any device. Search, copy, and paste code from your favorite technical books directly into your application.

The perks don't stop there, you can get exclusive access to discounts, newsletters, and great free content in your inbox daily

Follow these simple steps to get the benefits:

1. Scan the QR code or visit the link below

https://packt.link/free-ebook/9781804617373

1. Submit your proof of purchase
2. That's it! We'll send your free PDF and other benefits to your email directly